NATURAL HISTORY OF DELMARVA

DRAGONFLIES AND DAMSELFLIES

to Kurt & Heidi

Hal White

7/16/2011

NATURAL HISTORY OF DELMARVA DRAGONFLIES AND DAMSELFLIES

ESSAYS OF A LIFELONG OBSERVER

BY HAL WHITE

WITH PHOTOGRAPHS
BY THE AUTHOR
AND JIM WHITE

UNIVERSITY OF DELAWARE PRESS
IN COLLABORATION WITH THE
DELAWARE NATURE SOCIETY

ISBN 978-1-61149-000-8

∞ The paper use in this publication meets the requirements of the American Nation Standard for Permanence of Paper for Printed Library Materials Z39.48-1984.

Cataloging-in-Publication Data is on file with the Library of Congress.

Design and production by Jerry Kelly.

PRINTED IN CHINA

To the memory of George H. and Alice F. Beatty
whose influence goes far beyond introducing me to
dragonflies and damselflies.

&

To Bob Lake whose professional work
on the aquatic insects of Delaware laid the foundation
for my avocational explorations
on the Delmarva Peninsula.

CONTENTS

FAMILY VIGNETTE TITLE	SPECIES LATIN NAME	SPECIES ENGLISH NAME	VIGNETTE NO

PART I · *Dragonflies (Anisoptera)*

PART II · *Damselflies (Zygoptera)*

JEWELWINGS (CALOPTERIGIDAE)

SPREADWINGS (LESTIDAE)

POND DAMSELS (COENARGRIONIDAE)

*Species that have not yet been documented on the Delmarva Peninsula but are likely to be present.

PREFACE

When I first became interested in dragonflies about fifty years ago, there was a need for field guides and natural history books about dragonflies and damselflies (Odonata). However, with the flood of these books in recent years, each focusing on a different region, one can ask, why another? And, what is so special about the Delmarva Peninsula that it deserves attention? Both are legitimate questions and ones that delayed my commitment to this effort in the face of annual requests and prodding from Lorraine Fleming and Linda Stapleford of the Delaware Nature Society. I needed a vision for a book that would be somewhat different from the existing books and would complement, rather than compete, with them.

First and foremost, I wanted to write a book that would illuminate my passion for dragonflies and damselflies, an avocation that undoubtedly seems a bit odd to many people. In that effort, each of the more than 125 species known from the Delmarva Peninsula serves as a prompt for a story, reflection, or other short vignette that creates an eclectic mix that I hope will convey to others why I find these large and colorful insects so interesting. I hope this book will stimulate casual reading in ways not normally associated with field guides.

Second, I wanted to write for people who already had an interest in dragonflies and damselflies and wanted to know more. They may or may not have a field guide already, which could make a difference in how they read the book. Those having a field guide likely will go to specific vignettes to find out more about a species they have just seen. Those who do not have a field guide (and hopefully many of those who do, as well) may read the book from cover to cover, bit by bit, for general interest. As a teacher, I know the power of stories, and many vignettes tell stories that might be read, remembered, and retold by anyone with an interest in natural history, whether or not they have a Delmarva connection.

And third, using dragonflies and damselflies, I wanted to illustrate with a specific group of organisms biological principles that apply to all living things including humans. I have tried to weave together concepts of evolution and natural selection, life cycles, behavior, physiology and anatomy, ecology, and conservation. In addition I have tried to introduce ideas about how science is done and the people who do science. Thus, those who have less interest in the specifics of Odonata can relate information here to the world around them in other contexts and get a sense of how we know what we know. In this context, one of my daughters, a fourth grade teacher, suggested that this book could be used as a biology textbook for elementary school teachers.

To some extent, my vision for this book is inspired by a childhood story told by Richard Feynman, the late physicist, Nobel Prize winner, and investigator of the space shuttle Challenger disaster. His father warned him against just knowing the names of birds. He said, "You can know the name of a bird in all the languages of the world, but when you are finished, you will know absolutely nothing whatever about the bird. You'll only know about humans in different places, and what they call the bird. So let's look at the bird and see what its *doing*—that's what really counts (38)." Knowing names enables communication, but what is so interesting to me is what Feynman's father knew so well. It is what different organisms *do* that really counts. Identification is only the beginning of appreciation, not the end.

Natural History of Delmarva Dragonflies and Damselflies will help with identification, but it is not a full-fledged field guide. It has some characteristics of a field guide in that every species known from the Delmarva Peninsula is discussed and illustrated. However, typical field guides for dragonflies and damselflies start with introductory sections describing dragonfly morphology, anatomy, behavior, taxonomy, evolution, and so forth, followed by descriptions of every species in a well-organized, but terse, format that includes pictures, identifying characteristics, habitat preferences, seasonal distribution, and geographic distribution. That format is worth while, but it is not the format or style of this book. Please use those books to supplement this book. Among the excellent regional field guides that together include all of the species found on the Delmarva Peninsula, of particular note are: *Damselflies of the Northeast* (49), *Dragonflies and Damselflies of Georgia and the Southeast* (4), *A Field Guide to the Dragonflies and Damselflies of Massachusetts* (58), *Dragonflies and Damselflies of Northeast Ohio* (70), and *Field Guide to the Dragonflies and Damselflies of New Jersey* (3).

My goals are to illuminate and illustrate each species with an interesting story, brief essay, or anecdote that integrates much of the usual introductory material with photographs, species commentaries, and perhaps offer a broader lesson as well. The title of each vignette appears along with the Latin name and the English name of the species that inspired it. The vignettes in this book can be read both for pleasure and for information. Despite my deliberate disruption of traditional structure, my training in science and career as a teacher dictate that this should also be a scholarly work of sorts. Therefore, I have included text citations (italicized numbers within parentheses) linking to references listed at the end of the book for those who would like to follow up certain points. Internal cross-references to species or topics discussed elsewhere in the book are indicated by bold numbers within brackets.

The Table of Contents is constructed to accommodate three types of readers. Those who have little prior knowledge of dragonflies and damselflies can get an idea of the topics from the titles given to each vignette. Dragonfly specialists can look up species by their Latin scientific names. And non-specialists with some familiarity with these insects can identify species by their English or "common" names that were approved and updated by the Dragonfly Society of the Americas. The species appear in alphabetical order within family groups based on their scientific name. This taxonomic order parallels that found in most field guides.

While I have taken many of the photographs in this book, at least an equal number come from others who share my interest in natural history and Odonata. As a challenge and matter of principle, my goal was to limit the pictures used in this book to unposed photographs taken on the Delmarva Peninsula. The result falls short of that goal particularly for rare species and species that rarely perch. In those cases, photographs of hand-held individuals or photographs taken in nearby states were used. The names of photographers appear in small print beside the photographs they provided.

Since moving to Newark, Delaware, in 1971, I have concentrated on the dragonflies and damselflies of Delaware, seeing its political boundaries as defining of my territory. Unlike humans, however, insects and other organisms don't know about political boundaries. Their distribution is constrained by natural boundaries and thus it made sense to expand my horizons for the purposes of this book to the entire Delmarva Peninsula, which includes parts of the "territories" of other "Ode" enthusiasts. For years Richard Orr has maintained and annually updated a list of species for Maryland by county including the nine counties on Maryland's Eastern Shore. Similarly, Steve Roble keeps track of Virginia Odonata including those found on the two Virginia counties that form the southern tip of the Delmarva Peninsula. Richard and Steve have unselfishly shared their

The Delmarva Peninsula

records and knowledge so that this book could be more complete.

The Delmarva Peninsula is not considered an island; therefore, I had to decide, somewhat arbitrarily for the purposes of this book, where to draw a line that separates it from nearby regions to the north. I could have followed the commercial interests of the early 19th century who made that decision by digging the Chesapeake and Delaware Canal. It cuts across the Peninsula near its narrowest constriction, shortens the shipping route between Philadelphia and Baltimore by almost 300 miles, and avoids ocean travel. The narrowest distance between the tidal Chesapeake and Delaware bays was once only 12 miles; but now, with the canal, only a small part of the original peninsula in the north is not surrounded by water. Access to the Delmarva "Island" is limited to boats and seven bridges, five of which cross the C & D Canal. The C & D Canal now constitutes a human-made natural barrier for many organisms; however, it is not a physical or biological barrier for dragonflies and damselflies that routinely fly across water.

Another possible boundary to use for Odonata would be the Fall Line. It divides the Coastal Plain from the Piedmont region and defines a physiographic boundary that in turn represents a biological boundary. This line runs roughly along Interstate Highway 95, US Route 40, and the Amtrak railroad from Wilmington through Newark in Delaware to Elkton, Maryland. There are more than a handful of species found commonly on one side of the boundary, but infrequently or not at all on the other.

I have used neither the C & D Canal nor the Fall Line as the northern boundary for the Delmarva Peninsula. As a matter of convenience and in line with others (*97, 108*), I have used a political boundary (Pennsylvania) in drawing the line between the Delaware River and Susquehanna River to define the northern boundary of the Delmarva Peninsula (See map). It is not the same decision a dragonfly would make. However, it includes all of Delaware and all of Maryland east of the Susquehanna River. In addition, it captures the transition zone between the Coastal Plain and the Piedmont and thereby includes several species that otherwise would be left out if either of the other boundaries were selected.

For each species, I have included a small county map of the Delmarva Peninsula. The known distribution for a species is indicated by the counties colored in black.

Because odonates fly over water, one could reasonably ask whether the Delaware and Chesapeake bays constitute physical boundaries for dragonflies and damselflies. For many species, the bays are not barriers, but for others, particularly damselflies, the distributional evidence indicates significant barriers. For example, there are several species, sometimes locally common, found on the Coastal Plain in the Pine Barrens of New Jersey that have never or rarely been found in Delaware or farther south. In the case of the Chesapeake Bay, at least one species on the Delmarva Peninsula is morphologically distinguishable from its brethren south of the Chesapeake Bay.

On the more serious side, beyond the trivial issues of deciding where to draw boundaries, this book is a plea to humankind on behalf of a few of our fellow earthlings whose survival we threaten—not through deliberate actions against them, but through our relentless destruction and disruption of fragile and unique habitats. Our seemingly innocent routine activities of building houses, fertilizing lawns and crops, salting roads in the winter, cutting down stream-side vegetation, tapping ground water supplies, introducing alien plants, and many other activities will contribute far more to the demise of certain species than most people realize. I would be delighted, but highly surprised, if all the dragonflies and damselflies that occur on the Delmarva Peninsula now are still here a century from now. I am saddened by having a nationally designated "Wild and Scenic River" whose beauty is a veneer that hides the fact that it is a relatively lifeless skeleton of what it could and should be, and most certainly once was. I am hopeful that someday, through the protection provided, the full biological diversity will return to White Clay Creek. If these vignettes in any way motivate others to prevent the further loss or disruption of wetland and aquatic habitats, the effort will be wholly worthwhile.

HAL WHITE

Newark, Delaware, 13 July 2010

ACKNOWLEDGMENTS

I would never have initiated this book without the prodding a decade ago by Linda Stapleford and Lorraine M. Fleming of the Delaware Nature Society. I thank them for having the vision for a book on the Odonata of the Delmarva Peninsula and for encouraging me to undertake this project. Subsequently, Jim White (no relation) of the Delaware Nature Society has given tremendous support in several ways. He has provided me with transportation and numerous enjoyable hours in the field exploring Delmarva's diverse wetlands in the pursuit of hard-won photographs of elusive species. Many of his photographs grace this book. In addition, Jim and his wife Amy's experience as coauthors of *Amphibians and Reptiles of Delmarva* (97) has guided me in the completion of this book.

Mike Riska, Executive Director of the Delaware Nature Society, has provided his support and encouragement from the start. Mary Richards, as chair of the Society's Publications Committee and Professor Emerita of English at the University of Delaware, has served most importantly as a liaison between the Delaware Nature Society and the University of Delaware Press represented by Don Mell and Karen Druliner. Mary also provided the final editing of the manuscript. More recently Gregory Inskip, current chair of the Publications Committee, has guided this project to completion.

Numerous friends and associates have read early drafts and have given valuable comments and critiques. Each has contributed significantly to the readability and accuracy of the text. In addition to Linda Stapleford, Lorraine Fleming, and Amy White of the Delaware Nature Society, my wife Jean and daughter Laura (who created the image on title page and book binding) and technical experts Clark Shiffer, Richard Orr, Steve Roble, Roland Roth, Jim White, Christopher Heckscher, and Giff Beaton have provided critiques of various drafts. All have my thanks for their comments and helping me avoid embarrassing errors or inaccuracies. Any errors that may remain are my responsibility.

While Jim White and I took most of the photographs in the book, there were many species that we were unable to find and photograph in the field. Photographs of one or more of these species were graciously provided by a variety of dragonfly experts and nature photographers. They include Jim Brighton, Michael Blust, Rick Cheicante, Steve Collins, David Czaplak, Randy Emmitt, Jeff Gordon, Hans Holbrook, Bill Hubick, Dick Lighty, Bob Moul, Annette Oliveira, Richard Orr, Clark Shiffer, Joe Sebastiani, and Judy Semroc. Their names appear next to their photographs. Shelly McCoy skillfully used GIS data to produce the Delmarva map in the Preface. I thank Jerry Kelly whose years of experience in book publishing and artistic sense is revealed in his beautiful layout for this book.

A book with color photographs needs financial support. I thank Daniel Rich, Thomas Apple, and George Watson in the Office of the Provost and the College of Arts and Sciences at the University of Delaware, Howard Brokaw, the Ellice and Rosa McDonald Foundation, the Fair Play Foundation, and the Delaware Nature Society for making this book possible.

PART I DRAGONFLIES (ANISOPTERA)

1 ALONE IN A FAMILY
Tachopteryx thoreyi (Gray Petaltail)

It is not necessary to be an insect taxonomist to distinguish different insect orders. By grade school, children can tell a butterfly from a beetle, a bee, a fly, or a dragonfly. As one encounters more closely related organisms, distinctions become more subtle and less widely known. The insect order known as Odonata includes two major suborders, the dragonflies (Anisoptera) and the damselflies (Zygoptera). Those scientific names refer to differences in the shape of their wings. In dragonflies, the hind wing is wider than the front wing, particularly at the base, while in damselflies, the front and hind wings are rather similar in shape and are particularly narrow at the base. However, there are many other ways to tell the two groups apart without looking at the wing shapes.

Dragonflies are stronger fliers, tend to be larger and more robust, spread their wings wide when perched, and usually have a large round head with eyes close together. Damselflies are more slender and dainty, have a dumbbell-shaped head with widely separated eyes, and hold their wings together or only partly spread when perched.

The next major subdivision below order is family. For the first part of this book, we will focus on dragonflies, of which seven families occur on Delmarva. Of these, perhaps the most charismatic family is the Petaluridae, the petaltails, a dominant group of dragonflies at the time of the dinosaurs. There is only one species in eastern North America, the Gray Petaltail. It is one of only 11 species worldwide in the family and the only species not found around the edges of the Pacific Basin. To add to its allure, the Gray Petaltail has been seen on the Delmarva only once. That was near the Octoraro Creek in Cecil County, Maryland, in the late 1930s (*40*).

Despite its absence from the recent record, the Gray Petaltail quite possibly still occurs in our region because the local, specialized habitats it prefers still exist in this area. The larvae (also known as nymphs) of this large primitive dragonfly are semi-terrestrial in that they live on wet soil under leaves near permanent woodland hillside spring seeps (*33*), of which there are many in the piedmont regions of Cecil and New Castle counties. Adults stay near the larval habitat and have a habit of perching on tree trunks in the sun where their mottled gray bodies blend with the bark thus providing camouflage. Frequently, one's first awareness of a Gray Petaltail is when it suddenly appears out of nowhere and lands on a shirt or trousers, mistaking the person for a tree trunk.

The Harlequin Darner [12] is the only other dragonfly in Delmarva that is likely to do this.

In addition to being quite localized and living in places where few other dragonflies occur, the multiyear life cycle of the Gray Petaltail also may contribute to the species' elusiveness. Consider a recently colonized habitat. It may be several years before adults would emerge, yet the hard-to-find larvae would be present—somewhat like the underground larvae of the seventeen-year cicada, but with a shorter life cycle. Thus, an ideal habitat where Gray Petaltails have not been found needs to be revisited yearly. To find a species that has been missing for over 70 years on the Delmarva Peninsula would be an exciting and welcome discovery.

HAL WHITE

The Gray Petaltail typically perches flat on tree trunks where its mottled gray body provides camouflage. Unsuspecting humans may be surprised when one suddenly appears and lands on their shirt or pants. This male was photographed by the author in central Pennsylvania.

2 ANDROCHROMICITY
Aeshna tuberculifera (Black-tipped Darner)

People who want to observe dragonflies usually go to the aquatic habitats where dragonflies breed. Invariably, they see many more adult males than females, a pattern also reflected by specimens in museum collections (*62*). One might conclude that the sex ratio is skewed. However, that is not the case. The sex ratio of dragonfly larvae and the cast skins (exuviae) they leave behind when they emerge is always close to 50:50 (*25*). The apparent over-abundance of male dragonflies has a simple explanation. Males are much more conspicuous and abundant where the collectors go. Males set up, patrol, and protect territories at their breeding sites. When a female dragonfly comes to the water to lay eggs, males nearby will immediately pursue her, often before a human observer has a chance to see her. Because females spend most of their time away from water, they are more dispersed and difficult to find.

Another reason people see more males than females is due to sexual dimorphism, the phenomenon where males and females of the same species look different. In the case of dragonflies (and damselflies as well), males tend to be brightly colored and showy while the females often appear dull and drab. Dragonflies share this trait with many birds, for example, peacocks and many songbirds. Both dragonflies and birds have keen eyesight. The bright, showy males stake out territories and defend them against other males. Being seen and recognized is apparently important in their competition for mates.

When scientists observe unusual patterns such as perceived unbalanced sex ratios or sexual dimorphism, they generate hypotheses to explain them (*39*). Exceptions to the rule often provide useful hints. Among birds, there are species where males and females look rather similar—Canada Geese or Mourning Doves, for example. The absence of sexual dimorphism in birds often occurs in species that mate for life and in which males are not territorial. However, such behavior is unknown in dragonflies.

The Black-tipped Darner represents an exception to the pattern of sexual dimorphism among dragonflies. In related darners, sexual dimorphism is the rule with only a few females having the bright colors of the male. However, all female Black-tipped Darners are brightly colored like males (androchromic). Although black, both sexes have many small bright blue abdominal spots and two diagonal blue to pale blue-green stripes on the sides of the thorax. Females also behave more like males by flying conspicuously. Similar numbers of male and female Black-tipped Darners reside in collections. Assuming that humans with nets are typical of natural predators in this regard, it is hard to imagine what Black-tipped Darner females gain selectively by looking like males.

Although inland, Black-tipped Darners are known from the mountains of North Carolina, nearer the coast northern Delmarva is about as far south to expect them at lower elevations. The only Delmarva record of this species I collected in an old beaver meadow at Lums Pond State Park in early October 1982. A population existed in nearby Delaware County, Pennsylvania in 1945 at a location that is now part of suburban Philadelphia (5). As happens to many species living near their range limits, human activities are altering or destroying the few preferred habitats that can support populations of the Black-tipped Darner. Any observation of this species now on the Delmarva Peninsula would be noteworthy.

The brightly colored female Black-tipped Darner ovipositing above has a similar color pattern to the males (left) of this species. Note how the female, like damselflies and other darners but unlike other dragonflies, inserts her eggs into plant tissues. Photographs taken in central Pennsylvania.

3 DINING ON MIDGES
Aeshna umbrosa (Shadow Darner)

Although present throughout the summer, the Shadow Darner seems to attract attention most in the fall when it sometimes flies in open windows on warm days. When this happens in a classroom or busy office, the effect can be dramatic. Being strong fliers and almost three inches long, the Shadow Darner can move swiftly about a room, inadvertently frightening humans who may think it is a giant wasp and with a corresponding sting.

Normally, however, Shadow Darners prefer small woodland streams, where males patrolling for females fly low and follow the shoreline. They leave this habitat to feed on midges and other small insects that often fly in clearings or at the edges of fields protected from the wind. Sometimes, if the conditions are right, hundreds of feeding dragonflies slice back and forth through clouds of small insects—like sharks attacking and terrorizing a school of fish. Typically, darners and some emeralds [44-47] congregate in these swarms that usually form in the late afternoon and may persist until it is dark.

This male Shadow Darner is perching in the sun to get warm on an early November day. Like other darners, its large compound eyes, which meet along the midline of the head, enable it to see in almost every direction. Getting close for a picture requires patience and slow motion.

Because the Shadow Darner prefers a shady habitat and can feed at dusk, it is appropriately named. The fact that it can negotiate in shade and see the almost invisible midges it feeds on in dim light probably explains why it will fly in open windows. Perhaps the Shadow Darner has special adaptations for seeing in low light, but in every outward respect it has the same large eyes as other species and can see in every direction except directly backward, where the body is in the way. For those who wish to net a Shadow Darner, a well-timed, fast swing from the rear is more likely to be successful than a swing from any other direction.

Among the darners of the northern Delmarva Peninsula, the Shadow Darner is fairly common and one of the last dragonflies to disappear in the fall. Chances are that any large dragonfly seen in November even when the temperatures are in the mid 50s after a frosty morning is the Shadow Darner. In the hand, it has a regular pattern of small greenish yellow and blue spots on a slender, predominantly dark brown abdomen and is distinctive for having blue spots on the underside of abdominal segments 4 through 7. It has two yellow stripes with a hint of green on each side of its thorax, and the paddle-shaped appendages at the tip of the male's abdomen have a point projecting backward and downward when viewed from the side. The wings sometimes have a brownish tinge in older individuals.

4 LOCAL EXTINCTION

Aeshna verticalis (Green-striped Darner)

Dragonflies known as mosaic darners [2-4] impress us by their large size, aerial agility, and especially their beautiful bright blue markings. To the north, their abundance and species diversity increases as does the difficulty of telling one species from another. Across Canada and the northern United States during August and September, mosaic darners occur at virtually every wetland. Large swarms involving several species frequently congregate near dusk in woodland clearings to feed on midges and other small insects. Children with nets pursuing these three-inch-long giants of the insect world become big game hunters, often missing swing after swing.

Of the thirteen species of this genus known from New England, only three are known from the Delmarva Peninsula. Two of those, the Black-tipped Darner [2] and the Green-striped Darner, have been documented but once, both on October 3, 1982, in the same wet meadow at Lums Pond State Park in Delaware. Thus, with the exception of the Shadow Darner [3] in this group, mosaic darners are quite rare in our area.

While only one documented record for the Green-striped Darner exists on Delmarva, a small population of this or a related species [2] existed at Lums Pond. Several times between 1982 and 1997 I saw unidentified mosaic darners in the appropriate habitat that looked like Green-striped Darners. One hopes that they might return, but rapid residential growth and consequent degradation of water quality and habitats make that unlikely. Some say, "Don't worry, there are other populations." However, similar effects can be accruing in those populations as well. Global extinction does not occur all at once. Instead, it occurs one locality at a time. Some day the last small, local population becomes extinct. Then there is no possibility for recovery.

Even to the north, Green-striped Darners have selective habitat preferences. They occur locally in grassy wetlands where males stake out territories. They frequently hover above the grass as they patrol and occasionally drop down to near water level to inspect for females that may be ovipositing in stems there.

The Green-striped Darner differs from other mosaic darners in that the prominent stripes on the sides of the thorax are green rather than blue. In addition, the first stripe has a characteristic wiggle and a backward extension near the base of the wing. Like most other mosaic darners, male Green-striped Darners have pairs of bright blue spots on the top of the abdomen.

To our north, the Green-striped Darner frequents grassy wet meadows in late summer and fall. Its presence on the Delmarva Peninsula is limited to one confirmed record. This male was photographed in northeastern Ohio.

5 MIGRATING MOSQUITO HAWKS
Anax junius (Common Green Darner)

When one imagines a dragonfly, the Common Green Darner usually comes to mind. It is the first species to appear in the spring and one of the last to disappear in the fall. It is the only species known from all fifty states, and it occasionally crosses the Atlantic where it sporadically appears in England [6]. Like the first migrant robins, harbingers of spring in the northeastern United States, the first Green Darners at local ponds signal the beginning of a new dragonfly season and create a buzz of excitement on the Odonata ListServ in late March or early April.

It puzzles me that the common name for this familiar dragonfly is not "Blue Darner" because most of the visible part of the male abdomen is brilliant blue. The thorax is green and not so apparent while in flight. As one moves north, there are quite a few other darners that appear blue in flight, so I guess Green Darner is appropriate. Given the easy recognition and wide distribution, it certainly is appropriate that a children's book has been written about the Green Darner (66).

The annual north-south migration of the Green Darner has intrigued entomologists for quite a while. The story has yet to be worked out. Unlike the Monarch butterfly that must migrate south each year because it cannot survive the winter, the larvae of Green Darners do quite well in the north and can be found in ponds in the middle of winter throughout their range. However, the fully mature individuals that suddenly appear in the spring and lay their eggs in local ponds did not overwinter here. We do not know the source of Green Darners in the spring. Unlike the fall migration, which is well-documented, the spring migration is by deduction. With the exception of a lone contrary observation that Rudy Raff and I made many years ago in central Pennsylvania (98), evidence for early spring emergence at ponds in the north is lacking in the literature.

Usually starting in late August through early October, migratory Green Darners, the offspring of adults appearing in the early spring, emerge and often swarm over fields where they catch food in preparation for a flight south. These autumn Green Darners tend to be larger and purplish, rather than a mature blue color, and they do not engage in breeding behavior. Like migrating avian hawks, these migrating mosquito hawks wait for a strong cold front to provide winds to assist them in their flight south (1). Thus, there appear to be resident and migratory populations that are reproductively isolated from each other somewhat like the two populations of Canada Geese found on the Delmarva Peninsula.

The female of this tandem pair of Green Darners is ovipositing in submerged vegetation at a Pennsylvania pond. The male's abdominal appendages clasp the top and back of the female's head. In many species the female oviposits alone. The cast skin of a Green Darner larva is left behind when the larva crawls out of the water and metamorphoses into an adult dragonfly.

Some argue that the recent interest in dragonflies is in part due to birders who started counting migrating Green Darners at places like Cape May, New Jersey. Recently, Green Darner migration has been tracked by researchers in ultralight aircraft following Green Darners fitted with miniature radiotransmitters attached to their thorax (*102*). In three days one individual flew 112 miles from central New Jersey across Delaware to near Ocean City, Maryland.

6 KILLING FOR THE RECORD - VOUCHER SPECIMENS

Anax longipes (Comet Darner)

Is collecting specimens ethically and environmentally justified? Dragonflies, along with butterflies, reveal interesting aspects of human values. Both are large and sometimes spectacularly beautiful insects. Despite folk names like "horse stingers" or "devil's darning needles", people generally consider dragonflies as beneficial or at least harmless. This puts them in a class distinct from mosquitoes, horse flies, cockroaches, termites, ants, and wasps. Many people, who would not give a second thought to stepping on an ant or slapping a mosquito, find the deliberate killing of a dragonfly for collection and study objectionable. It is sobering to think that the number of dragonflies killed incidentally by motor vehicles or through habitat destruction exceeds by far the number killed deliberately by collectors [65].

In 1999 a world authority on dragonflies, the late Philip Corbet, was physically threatened when he caught a Common Green Darner [5] from North America in southwestern England and planned to keep it as a "voucher specimen" (24). This is the term used among specialists for specimens that document the presence of a species in case someone might wish to question or confirm an observation at a later date. Sight records are notoriously unreliable and photographs frequently prove inadequate, so nothing beats a specimen for examination. Nevertheless, the transatlantic Green Darner was released but not without a frightening human confrontation that provoked an international firestorm on the Internet.

I began collecting dragonflies as a teenager. It was a perfectly respectable hobby that laid the foundations of my career in science. It established certain principles of observation, record keeping, curation, and documentation tied directly to specimens. I still collect dragonflies, but much more selectively. Even then I am challenged by those close to me to justify collecting. For most people, a dead specimen has no value compared to a beautiful living dragonfly. I know of no specialist who does not also marvel at the beauty of a living dragonfly. For me now, to kill a dragonfly entails a conscious decision that to do so has potential scientific value. It carries with it an obligation to preserve the specimen for future use.

While I possess many specimens that have never been studied closely, I also have many that have proved to be very useful in recognizing new species and defining geographic ranges of species and subspecies. I also have specimens that provide the only documentation that certain species occur on the Delmarva Peninsula. Ironically, such specimens aid conservation. For example, how can a species be declared and defended as an endangered species if it can't be identified, described, and verified?

I associate the idea of voucher specimens with Comet Darners. One day I saw one of these spectacular and unmistakable large dragonflies with a green thorax and a bright red abdomen flying out over a pond in central Pennsylvania where it had not been documented previously. Because Comet Darners are uncommon and difficult to catch, they appear infrequently in collections. My announcement that I had seen this species was met with skepticism and the offer of $10 for a specimen. In 1959, $10 could buy a lot more than it does now and

was a handsome sum for a high school bounty hunter. Considerable patience, some luck, and a well-coordinated swing produced a voucher specimen of tangible value. That voucher specimen now resides in a collection at Penn State University. I have only three voucher specimens of the Comet Darner to document the species from the Delmarva Peninsula.

Comet Darners prefer flooded sand pits and other ponds—preferably fishless. They are most often seen in June and July. While the abdomen of the male is orange-red, the female abdomen is duller reddish with pale spots. Their long orange-brown and black hind legs also are distinctive.

The male Comet Darner (above and below) is large, striking in color, and has distinctive long legs. It is one of the few dragonflies that can be identified 50 yards away as it courses over a pond. In flight the long legs are held close to the body.

7 CHANGING ETHICS
Basiaeschna janata (Springtime Darner)

In rural central Pennsylvania where I grew up, the first day of hunting season in the fall and the first day of fishing season in the spring were unofficial holidays for most boys, starting in elementary school. My contemporaries dreamed of shooting a buck or catching a monster trout. I did not. Having moved to the area from elsewhere with my parents, I lacked a local family history of farming and living off the land. Although I lived next to a renowned trout stream, I found other pursuits more interesting than fishing and hunting—activities like looking for arrowheads or catching crayfish and tadpoles.

No one questioned killing a rabbit, deer, pheasant, or trout so long as it was used for food and not wasted. Killing a robin with a pellet gun, however, was taboo, but not particularly so for grackles or starlings. A kid collecting insects was a bit strange, but innocent enough. Curious people would ask, "Whach ya catchin?" The answer would be followed by, "Whach ya do with em?", as if judgment were based on utility rather than a prohibition of killing.

HAL WHITE

The Springtime Darner, our smallest darner, flies low to the water over small streams in April and May.

Times have changed. Most people grow up in urban environments and are quite removed from their source of food. "Thou shalt not kill" is applied rather frequently as an absolute unmodified by purpose. In recent years, interest in dragonflies and damselflies has grown enormously. As a consequence, an interest once dominated by collectors is shared by an even larger number of people who find collecting unappealing, even reprehensible. The new ethic is that these magnificent, large, colorful insects should be observed and perhaps photographed, but not collected. I only wish that wetland habitats now overrun or altered by suburbia could be treated similarly. The prohibition against killing applies to the deliberate personal act, while clearing land and building houses impersonally leads to relatively acceptable massive killing.

The Springtime Darner is an example of a species far more threatened by urbanization than by collectors. The streams across the Delmarva Peninsula provide ideal habitat for the Springtime Darner, our smallest darner. True to its name, it flies in the spring patrolling stream edges starting shortly after the opening of trout season in April. Based on larval sampling, the Springtime Darner inhabits many local streams, though it is not often seen as an adult. It is hard to imagine that collecting adults would noticeably diminish its populations. However, the inexorable urbanization presents a huge threat to the long-term survival of this lovely, relatively small darner with its blue abdominal spots on each segment and yellow stripes on the sides of its thorax. Females have a similar appearance but the abdominal spots may be greenish. Its demise in urbanizing areas will likely go unnoticed by developers, planners, and the local human population.

8 HORSE STINGERS AND OTHER FOLK NAMES

Boyeria vinosa (Fawn Darner)

Some folk names for dragonflies conjure up vivid images. Where would names like "devil's darning needle" and "horse stinger" come from except from people who had an aversion to insects? After all, if small colorful insects like bees and hornets can nail you with their sting, imagine what something the size of a dragonfly could do!

People who have made a hobby of learning the colloquial names for dragonflies in different parts of the world see interesting cultural patterns (52). European names, like those above, tend to emphasize allegiance with snakes or the devil. For example, in Spanish the folk name for dragonfly is *caballito del diablo* (little horse of the devil). A few folk names like "mosquito hawk" show more understanding. In Japan, dragonflies (*tombo*) are revered rather than feared.

Despite the presumption that a name like "horse stinger" has little basis in fact, an encounter I had with a female Fawn Darner in Massachusetts suggests otherwise. Technically speaking, an insect sting (as opposed to a bite) comes from the ovipositor. Therefore, only female insects can sting. It was late on an August afternoon when I decided to take a break and sit down on a rock with my feet in the water of a fast-moving stream. Shortly thereafter I felt a sharp pain in my right calf and instinctively slapped what I expected to be a wasp or large horsefly. Much to my surprise, it was a female Fawn Darner that had mistaken my leg for a streamside log in which to lay her eggs. I know of two other people who have had a similar once-in-a-lifetime experience of actually being stung by a dragonfly. Certainly such an experience by someone less familiar with dragonflies could be the source of a common name like "horse stinger". However, despite my singular experience, they are not to be feared. Furthermore, among dragonflies, only the darners have ovipositors capable of a sting.

The Fawn Darner is a fairly common stream dragonfly in late summer and early fall. Males patrol the shoreline just above the water surface searching for females around rocks and under overhanging tree roots. Because they often fly in a shady environment, they will fly on cloudy days and feed at twilight. They are moderate-sized, brown dragonflies that derive their name from their color. Males and females look pretty much alike. Both have two bright yellow spots on each side of the thorax and small yellow spots down the side of the abdomen. The larvae of the Fawn Darner are interesting because they play 'possum when handled.

The female Fawn Darner is ovipositing in wet wood (not in flesh) along a stream in Pennsylvania. In general appearance, the male and female Fawn Darner look quite similar and do not exhibit the sexual dimorphism often found between males and females of other species.

9 HOMELAND SECURITY
Coryphaeschna ingens (Regal Darner)

At the extreme southern tip of the Delmarva Peninsula is a speck of land, Fisherman Island National Wildlife Refuge. This place marks the point of departure or arrival for those who drive across the 17.6-mile Chesapeake Bay Bridge-Tunnel. As with Cape May, New Jersey, this southern tip of land is also a point of departure and arrival for migrating birds. Consequently, it has become a favorite birding spot. That was until 2004 when the Department of Homeland Security declared it off limits. Apparently the authorities felt it was easier to restrict access to the island than to worry about suspicious characters wandering around with telescopes and binoculars, possibly plotting to destroy the bridge. Birders complained. Access to the island is now only by permit.

The fact that many birders have also become interested in butterflies and dragonflies paid off in this instance. Since 2001 they have reported not one, but two, species of southern dragonfly from Fisherman Island that are otherwise unknown from the Delmarva Peninsula. They are the Regal Darner and the Roseate Skimmer [70].

On July 26, 2003, Larry Brindza was participating in a butterfly count sponsored by the North American Butterfly Association when he saw a Regal Darner perched on a tree trunk by a marsh on the north side of Fisherman Island. He was able to watch the dragonfly at close range with binoculars and noted in particular the vivid blue eyes characteristic of the female of this species. (Young female Regal Darners have green eyes like the males.) Normally, such a sight record would be questioned without photographic documentation or a voucher specimen; however, Larry easily distinguished it from a Swamp Darner [10], the only other reasonable possibility. Thus, the record is virtually certain but not confirmed with a voucher specimen or photograph.

The impressively large Regal Darner (up to 4 inches long) with rather short legs is common in the southeastern United States where it inhabits lakes, ditches, and slow streams. It has a long, mostly black abdomen with transverse green stripes and a green thorax with a wide brown stripe. It sometimes feeds on honey bees around a hive, giving it the name Bee Killer.

The sighting of a Regal
Darner at the extreme
tip of the Delmarva
Peninsula, far from
its normal range, was
undoubtedly a chance
observation. This one
was photographed in
Florida.

10 TRAPPING DRAGONFLIES
Epiaeschna heros (Swamp Darner)

We actually have a lot to learn about dragonfly behavior. That is why beginners can make interesting new observations and significant contributions to our knowledge. Here is a chance observation that might be the basis for a good high school science project.

Paul Schaeffer, an entomologist working for the USDA Beneficial Insects Laboratory in Newark, Delaware, has a wonderful curiosity. He seems to notice things that others miss. One day Paul appeared at my front door with three large dragonflies he wanted me to identify. All were male Swamp Darners. He discovered them in traps he had set for caterpillar-eating beetles. He constructed the simple traps from 20-oz plastic beer cups with pieces of cut plastic stapled to form a peaked roof over the opening (See picture). In May 1995, he attached a couple hundred of these traps to tree trunks in two woodlots heavily infested with Gypsy Moth caterpillars near Sudlersville in Queen Anne's County, Maryland.

Darners like this female Swamp Darner and other large dragonflies perch on the underside or side of objects with their bodies hanging vertically. (Skimmers and most Clubtails, in contrast, perch on top of objects with the abdomen held horizontally). To the right is one of the plastic beer-cup traps used to monitor gypsy moth predators that unexpectedly also attracted male Swamp Darners.

The trap was designed to work as follows. Beetles climbing up a tree encounter a plastic strip encircling the trunk that they can not cross. They then walk around the trunk seeking a path upward and sooner or later encounter a slot that goes through an opening from which they fall into the cup where Paul would find them. Paul continued his beetle survey and continued to catch male Swamp Darners in his traps, a total of ten in all (72). Once he found three in one trap, and another time there were two in one trap. Due to the dimensions of the trap, the Swamp Darners could only have entered from the front directly under the peaked roof.

Why only males? Why only Swamp Darners, one of the largest dragonflies found on the Delmarva Peninsula? What about the trap's design attracts them? Could this trap be adapted to attract and catch other kinds of dragonflies? We do not yet have answers to these questions.

As suggested by their common name, Swamp Darners live in swampy places. Typically these are woodland swamps that may dry up for part of the year. Females lay their eggs in wet, rotten wood in these swamps. Unlike other darners, male Swamp Darners do not set up and patrol territories, so we know little about their mating behavior, which could be related to the fact that only males were trapped in the story above. In drought years in the southeastern United States, Swamp Darners are more common in our area, presumably because they migrate or disperse to find suitable breeding places. Although Swamp Darners occur throughout the Delmarva Peninsula, a good place to observe them is along woodland roads and clearings near the Pocomoke River. Here they are hard to miss as they swoop around feeding on small insects in June and July.

11 FLYING HIGH
Gomphaeschna antilope (Taper-tailed Darner)

We know relatively little about the Taper-tailed Darner. What we do know is puzzling. For instance, this species is rare in collections and those specimens that end up in museum collections are often not collected by dragonfly specialists. In an article on the dragonflies of eastern Pennsylvania (7), the authors write about the Taper-tailed Darner: "It has long been known for its inclination to enter buildings and insect light-traps, but is seldom collected otherwise. In Philadelphia this species was collected on the 9th floor of the Flanders Building, the 14th floor of 1608 Walnut Street (3 specimens on different dates), and a male flew into the Academy of Natural Sciences." It is also known from buildings in Washington, DC, including the National Museum of Natural History.

With the advent of air conditioning, the windows of tall city buildings remain closed year round. Thus, records of the Taper-tailed Darner have tapered off because they no longer can enter center-city buildings at the 14th floor. It also may be true that because of air conditioning, humans remain inside for personal comfort and fewer people are outside in the places where the Taper-tailed Darner normally lives.

I have observed the Taper-tailed Darner only twice, once in southern Delaware and once in the New Jersey Pine Barrens. The only record for Delaware before I found it on June 15, 1975 near the Cypress Swamp in Gumboro, Delaware, was from the Ship John Light in the middle of Delaware Bay on June 1, 1936. More recently, about 20 were seen by Michael Blust as they were feeding with Swamp Darners [10] over a cornfield in Selbyville, Delaware, on July 1, 2005. Thus it appears that a population probably persists in the headwaters of the Pocomoke River far from tall buildings and central air conditioning.

In recent years, several Taper-tailed Darners were discovered flying along the coast north of Boston, Massachusetts, many miles from the swamps and bogs of southeastern United States to possibly southern New England where they normally breed. It appears that they disperse after emerging, which would explain their flying into city buildings and over salt water. The Taper-tailed Darner has a long slender body and is brown with little patterning. Compared to its cousin the Harlequin Darner [12], the cells in its wings are larger and fewer in number. Like the Harlequin Darner it perches vertically on tree trunks; however, the Taper-tailed Darner has a later flight season.

Taper-tailed Darners are most often found in June and July in the Pocomoke River watershed.

12 TRIBUTE TO DAVE NYE (1941-2002)

Gomphaeschna furcillata (Harlequin Darner)

Of all the people I have ever known, Dave Nye has to be the most unusual. My first encounter occurred in May 1971 in Massachusetts, when he appeared unannounced at my door on a day I had stayed home bedridden with a fever. If I had not been sick, I would have missed him and never known it. There he stood, paralyzed on his left side, dressed in army fatigues, wearing combat boots, a safari hat, and belt adorned with hand tools. When he came in, I soon learned that he was defiantly self-sufficient and living on full disability pay. He lived out of his specially modified Ford Bronco as if he were still in the Marines serving combat duty that precluded changing clothes and bathing. He relished eating moldy cheese because it didn't get wasted. Having cheated death by surviving a fight in which his head was laid open with a trenching tool at Guantanamo, he figured he was living on borrowed time, could take risks, and could ignore personal hardship. Having little sympathy for a softy graduate student who stayed home with a fever, he shamed me into taking him into the field to look for dragonflies, one of his numerous natural history passions.

Dave spent most of his time outdoors, was a keen observer, and had an incredible knowledge of things natural. Unrestrained by home, finances, or family, he roamed the country and dropped in unannounced at the offices of authorities on whatever interested him. When I by chance moved in 1971 to Newark, Delaware, Dave's nominal home address, he introduced me to such dragonfly hotspots of Delmarva as the Cypress Swamp. He insisted on leaving at 4:30 AM. Despite his paralysis, he drove, while telling tales of his many field experiences. Then he led the way through a thigh-deep, cold stream and greenbrier thickets to show me where he was the first person to discover the larva of the Harlequin Darner described by Sid Dunkle (*32*).

Several years later Dave gleefully showed me a progression of snapshots of his right hand necrosing from a Cottonmouth bite that he had received in Georgia. He refused to have it treated because he wanted to see what would happen. That he lived to 60 was a miracle. His passing as the result of an automobile accident in Alabama (*60*) brought me sadness. With him went a wealth of observations. As they say in the Pular language, "When an old person dies, it is like a library that burns." When I see a Harlequin Darner, I think of Dave and all the knowledge about the lives of dragonflies and other things that was lost with him when he died.

As often as not, I discover a Harlequin Darner after it has discovered me. Males like to hover in sunlit forest glades in and around sphagnum swamps and perch vertically on nearby tree trunks. A person standing in the middle of such a glade looks a lot like a tree trunk, at least Harlequin Darners seem to think so. They land on shirts, pants, and nets, and can be caught with a slow steady hand. From several feet away, their slender bodies have a dark greenish appearance. However, close up, they are a spectacular mottled mix of green, orange, and black that befits their common name.

Harlequin Darners can be abundant in April in the Cypress Swamp and Pocomoke drainage near the Delaware-Maryland state line, where recently emerged adults forage along woodland roads. They also occur regularly in the Nanticoke Wildlife Area and the Choptank River floodplain. Finding Harlequin Darners on the Delmarva Peninsula is always a sign of interesting, undisturbed habitats nearby.

Harlequin Darners often land on sunlit, vertical surfaces like tree trunks, but pants or shirts will do.

13 CAUGHT OUT OF BOUNDS
Gynacantha nervosa (Twilight Darner)

The Twilight Darner normally lives and flies year-round in Florida and is known from southern Georgia and Alabama and south to Bolivia. There is one disjunct record from Oklahoma. It is not known from the Carolinas or Virginia and certainly was not among the species expected from the Delmarva Peninsula. However, while examining the University of Delaware insect collection in 2006, I encountered a pinned Twilight Darner collected September 27, 1975, in Newark, Delaware. The collector listed on the label was Charles Mason, then a newly-arrived assistant professor of entomology at the university. Although after these many years he does not remember all the details from collecting the female Twilight Darner, he keeps field notes and knows what he did on that Saturday. The dragonfly was observed during early evening flying up and down a small stream located just behind his yard in Brookside Park in southeast

How this female Twilight Darner made its way to Newark, Delaware in September 1975 is a mystery because no other records exist north of Georgia. Among dragonflies and damselflies, the Twilight Darner must certainly rank among the least colorful. The male (bottom), photographed in the deep south, displays an almost patternless light brown body that makes it difficult to see it in its shady haunts.

HAL WHITE

GIFF BEATON

Newark. The specimen was captured with an aerial net as it passed upstream while the collector was partially hidden behind some tall vegetation growing along the stream bank.

How a Twilight Darner got from its normal southern home to northern Delaware is a mystery. Certainly, there are many examples of southern species carried north by winds from tropical storms in the summer and fall. The fact that the Twilight Darner also occurs in the Bahamas and West Indies indicates that it can disperse over hundreds of miles of ocean. In addition, several species of dragonflies routinely migrate north along the East Coast of the United States each year, but the Twilight Darner is not known to be one of them.

I am reminded of the time I collected a Zebra Longwing, the state butterfly of Florida, in Newark, Delaware. That species is a weak flier and feeds only on passion flower, a plant that grows in South Carolina, but not anywhere near Newark except as an ornamental. After I learned that the Zebra Longwing and other colorful species are shipped around the country as novelties to be released at wedding parties, I became less excited about my voucher specimen. So far, however, I have never heard of dragonflies being released at weddings, and the Twilight Darner is among our less colorful species, so its presence cannot be explained that way. Who knows? Perhaps it flew into a rail car or truck and got a free ride north.

On the off chance that another Twilight Darner will fly this far north, the following description may help. As described by Dunkle (34), it is a "slender, plain brown and dull green dragonfly" about three inches long. It has small legs and large eyes, which presumably enable it to fly and feed in the darkness at dusk. It breeds in fishless temporary pools. Southward it may be encountered in shady hummocks where during the day it "hangs out" in palmetto or woody vegetation, from which it may be flushed, only to perch again after a short quick flight.

14 DRAGONFLY PRESERVATION IN MUSEUMS

Nasiaeschna pentacantha (Cyrano Darner)

The insect collection at the Academy of Natural Sciences in nearby Philadelphia contains millions of specimens including many dragonflies and damselflies collected over a century ago by people like Philip P. Calvert. At that time and through much of the 20th century, entomologists carefully mounted specimens on pins, let them dry, and kept them with related specimens in sealed boxes containing moth balls, either *para*-dichlorobenzene (PDB) or naphthalene, to keep away other insects that might destroy them. Associated with each specimen on the same pin were one or two small labels with exceedingly small type on which were printed the species name, where and when it was collected, and the name of the collector—tombstones in an insect graveyard. In days gone by, the type was set and the labels printed in the museums on small hand-operated printing presses. It takes a lot of patience and careful work to prepare a museum specimen properly, so that it will be scientifically useful for centuries.

Interestingly, the standard method of preserving dragonflies and damselflies has changed since about 1960 in ways that save space, keep body parts together when they break, and avoid the need for a magnifying glass to read labels (6). Instead of being pinned with their wings spread apart, specimens are dried with their wings together and placed in clear plastic envelopes along with a 3- by 5-inch index card on which much information can be printed. The specimens are then placed in filing-card boxes such that they can easily be sorted and examined. Modern computers and laser-jet printers have made obsolete the small, hand-operated Kelsey printing presses formerly used for printing labels.

Current methods differ in other ways from the days of Calvert. Rather than deadly cyanide killing jars, specimens are killed in acetone and then soaked in acetone for several hours to a day depending on the size. The acetone does several things. It kills gut bacteria that normally would partially decompose specimens drying slowly in air. It extracts fats that hinder drying, and it partially dehydrates the specimen so that it dries quickly in air. An added benefit of this procedure is that it does a much better job of preserving colors and color patterns. One of the great disappointments of examining old, pinned specimens is that most are dull black, brown, or gray with little hint of the true beautiful life colors. While acetone has definite advantages and is not poisonous like cyanide, it is hazardous and should be used only in well ventilated spaces.

The Cyrano Darner specimen, shown in a plastic envelope (see picture), is

Nasiaeschna pentacantha (Rambur, 1842)   1♂
reconstructed wetlands by Howell School Rd, Lums Pond State Park,
Kirkwood, New Castle County, **DELAWARE**
39°34.21'N, 75°44.17'W, elev. ~60 feet
Saint Georges 7.5' USGS Quadrangle
24 June 2000   Coll. & Det. by H. B. White   Coll. No. 00-08

Museum specimens of dragonflies are no longer mounted on pins with tiny labels. Now specimens are placed in Mylar® or cellophane envelopes with data printed on 3- by 5-inch cards as shown for this Cyrano Darner. The slightly arched and tapered abdomen and prominent "nose" are a distinctive characteristic. Note how the vibrant colors of a live Cyrano Darner are lost in the preserved specimen.

named for its prominent and distinctive nose, like that of Cyrano de Bergerac of literature fame. Males of this rather large dragonfly typically take up territories in sunny glades along slow-moving, wooded streams or in coves along the wooded shores of ponds. Cyrano Darners fly slowly back and forth about four feet above the water. Their gray-green mottled abdomens have a noticeable arch and taper to the end. They are most common in June and July. Cyrano Darners have an unusual foraging behavior—something that would never be deduced from a museum specimen in a plastic envelope. In contrast to most other darners that course back and forth in open areas capturing small insects that happen to fly by, Cyrano Darners hover near leafy vegetation and look for perched prey that will fly up when disturbed and be caught. They also are known to catch and eat other species of dragonflies.

15 FARM POND CLUBTAIL

Arigomphus villosipes (Unicorn Clubtail)

Farm ponds are a great place to observe a variety of common dragonflies and damselflies at close range. Generally, the skimmers **[48-84]** are conspicuous, abundant, and quite cooperative subjects for photo opportunities. Males often find perches along the shore where they can defend their territories. Not quite so conspicuous nor as common is the Unicorn Clubtail, a light olive-green species with brown markings on the thorax that become black on the abdomen. From above, the tip of the male's abdomen looks like yellow pinchers. Males perch on muddy shores, spatterdock leaves, rocks, and flat surfaces within a few inches of the water surface. They are quite wary and fly directly to another perch when disturbed. Pursuing them in hopes of a good photograph can be frustrating.

While clubtails reach their greatest diversity in rivers and streams, the Unicorn Clubtail is atypical among clubtails in this area in that it prefers ponds and slow, mud-bottomed streams. Most other clubtail larvae require well-aerated flowing water, but the larvae of the Unicorn Clubtail seem to do well burrowing in mud. The elongated tip of the abdomen serves as a siphon to bathe the gills within the abdomen with aerated water. When the adults emerge, their abdomens are still touching the water, so it is important that they emerge when the water is calm. Otherwise, they might be washed away and damaged before their wings expand and harden.

The males of some clubtail species have cobra-like, enlarged ends to their abdomens (a clubbed tail), but most, including the Unicorn Clubtail, are less spectacular in this respect. The "unicorn" name comes from the small horn that projects from the surface between their eyes. In contrast to skimmers and darners, which have rather spherical heads with eyes that meet in the middle, eyes of the clubtails do not meet; and thus they have more of a dumbbell-shaped head, though not so pronounced as in damselflies.

Unicorn Clubtails emerge in May and persist into July and occasionally early August. With their long flight season and accessible habitat, it is probably the easiest clubtail to find on the Delmarva Peninsula, though it never occurs in the abundance of some of the stream clubtails that have synchronized emergence.

The Unicorn Clubtail often perches on flat surfaces and vegetation near pond edges (top) and vigorously defends its territory from other Unicorn Clubtails. Below, a Unicorn Clubtail shares a stump in peaceful coexistence with a Common Whitetail [75] of the skimmer family.

16 GETTING WARM...KEEPING COOL
Dromogomphus spinosus (Black-shouldered Spinyleg)

Dragonflies are poikilothermic. That is the scientific term for cold-blooded, which means their body temperature fluctuates with the air temperature and is modified by metabolic and behavioral factors. Some insects, like the bumble bees that we see flying around azaleas on 40-degree spring mornings or night-flying sphinx moths on cool nights, are large and compact with insulated bodies. They have physiological adaptations that generate heat and retain it before they fly.

Dragonflies have little insulation, and their sleek form radiates or absorbs heat readily from the environment. This explains in part why they are sun-lovers. (Don't spend much time looking for them on cool or cloudy days.) Nevertheless, it is common for dragonflies to "warm up" before they take off by vibrating their wings.

In the summer when the sun is high and the days are hot, sun-loving dragonflies can get overheated. In particular, I remember a blistering hot day of over 100°F in the Sand Hills of north central Nebraska, when all the dragonflies seemed to disappear. As an over-heated homeotherm, I took refuge in a copse of cottonwood trees and discovered thousands of dragonflies taking a midday siesta in the shade. In places they perched cheek-to-jowl on bare twigs. While this behavior occurs in the East, it is observed infrequently because shady perches are abundant and dispersed, and the temperature rarely exceeds 100°F.

To avoid overheating on hot sunny days, the Black-shouldered Spinyleg points its abdomen towards the sky in the obelisk posture (left) to expose less of its body to the sun's rays. Note the smaller shadow. The hind legs have prominent spines.

Dragonflies, more than damselflies, regulate body temperature by varying their orientation to the sun. On cool, sunny days and in the morning, dragonflies will congregate in wind-protected spots and perch, like sun-bathers, with their bodies and wings perpendicular to the rays from the sun so that they can absorb the maximum sunlight. On the other hand, on calm, hot days before retreating to the shade, dragonflies like the Black-shouldered Spinyleg assume the obelisk posture, pointing their abdomens directly at the sun to minimize heating (as shown by the minimal shadow cast by their bodies.)

The Black-shouldered Spinyleg occurs regularly, but uncommonly, on the Brandywine River and other large streams, including the Choptank River, from late May to early September. Males typically perch near the shore in the sun on logs and rocks in the water throughout the summer. Females and young adults perch on vegetation in woodland clearings near streams. They are often wary and hard to approach. Their prominent leg spines, which give rise to the common name, help them capture and hold prey. Occasionally this species will fly out and hit the surface of the water, possibly to get a drink [38] or as another source of keeping cool—evaporative cooling in this case.

17 DESCRIBING NEW SPECIES
Gomphus apomyius (Banner Clubtail)

Dragonfly specialists exploring the tropics of South and Central America can expect to encounter undescribed species. By contrast, we know the dragonfly fauna of the northeastern United States rather well, and only a few new species have turned up in the past half century. The Banner Clubtail is one of them and shows that there still may be discoveries to be made in well-explored areas.

In the mid 1950's, George Beatty discovered a new clubtail species in the New Jersey Pine Barrens and had planned to describe it formally. However, he procrastinated and, in the jargon of science, "got scooped" by Thomas Donnelly, who described the Banner Clubtail in 1966 after discovering it more than 1000 miles away in eastern Texas (*26*).

Considerable excitement typically accompanies the discovery of a new species, and there is a certain honor that goes with describing a new species in the literature. In general, discoverers share and compare their observations with other specialists before publishing. The discoverer has the option of letting someone else describe the species, which sometimes results in the species being named after the discoverer. This did not happen with the Banner Clubtail.

This dainty and pretty little clubtail is rare and local throughout its range. It occurs along sandy Coastal Plain streams where there is a significant current. The Banner Clubtail gets its common name from the bright yellow spot near the widened end of its stubby abdomen, which is held high as it flies and hovers over riffles. It has endangered status in New Jersey and probably should be listed as endangered on the Delmarva Peninsula, where I first observed it with Kitt Heckscher in 2004 on a tributary of the Nanticoke River in Sussex County, Delaware. Its flight season is brief and early, peaking in mid-May. The rapid and seemingly uncontrolled construction of housing developments in Sussex County could threaten this population in the near future.

Since the description of the Banner Clubtail, a few other species that occur on the Delmarva Peninsula have been described. However, these were cryptic species, meaning that taxonomists decided that something formerly thought to be one species should be split into two. This was the case for the Brown Spiketail [29] and Jane's Meadowhawk [78], described by Frank Louis Carle in 1983 and 1993, respectively (*21, 22*). The Banner Clubtail is not a case of splitting. It is distinctive and easily distinguished from related species.

A recently emerged (teneral) male Banner Clubtail (top) rests on the banks of the Nanticoke River in Sussex County, Delaware. After their wings have fully expanded, but not fully hardened, all dragonflies hold their wings together above their body in a position normally associated with damselflies. The mature male below displays his clubbed tail with its yellow banner.

18 TSUNAMI ALERT
Gomphus exilis (Lancet Clubtail)

The sun has barely burned through the early morning haze and reflects off the mirror-like water surface. From around the bend, the pleasant purr of a motor-boat gets louder. The boat passes and disappears around the next bend leaving the observer to watch in horror as its mesmerizing wake moves across the surface and laps against the muddy shore nearby. Inches above the waterline, thousands of emerging clubtails are now doomed—destined to be food for birds and fish, for they will never fly. Unknowingly, the boater, on an exhilarating early morning ride, has destroyed an estimated 32% of the population emerging that morning (53). The New York Times did not report this disaster that occurred on the Connecticut River in Massachusetts in 2003.

Similar events probably happen frequently in many places and go unnoticed. The timing is critical. Synchronized mass emergence is fairly common among clubtails. Although it provides safety in numbers against predators when the freshly emerged adults (tenerals) are fragile and most vulnerable, the behavior evolved in the absence of motorboats and other watercraft. Human activities affecting a couple of hours a year can have profound effects on certain dragonfly populations.

Like the Cobra Clubtails (*Gomphus vastus*) decimated above, the Lancet Clubtail emerges on muddy banks close to the waterline. It takes about an hour until it is ready to fly away. Although emergence is not limited to a single day, the bulk of the population emerges over a week or two in April and May. Fortunately for Delmarva populations, their preferred slow stream habitats are not favorable for motorboats. However, it is easy to imagine that an ill-timed thunderstorm and downpour could cause significant mortality.

Lancet Clubtails are our most common clubtails and, at less than two inches, one of our smaller clubtails. They are brownish with a yellow stripe extending the full length of the top of their abdomen. There is not much of a club in this clubtail. The Ashy Clubtail [20] often occurs with the Lancet Clubtail. However, the Ashy Clubtail is slightly larger, mostly brown, and lacks any bright yellow markings. The first Lancet Clubtails emerge in late April; and few remain at the beginning of July.

JIM WHITE

A pair of mating Lancet Clubtails (right) perched on a streamside leaf in Glasgow, Delaware. Note how the terminal appendages of the male's abdomen clasp the top of the female's head. A male Lancet Clubtail is above.

HAL WHITE

19 EXTENDING HABITAT AND GEOGRAPHIC RANGES

Gomphus fraternus (Midland Clubtail)

Clubtails are most common and diverse on unpolluted, clean-flowing inland rivers and streams. Only a few species prefer ponds or slow-moving water. Thus, the list of clubtails for the Delmarva Peninsula is relatively short. Several species are associated with the northern Piedmont, where rocky streams with good current exist. It was a total surprise in mid-May 2002 when Christopher "Kitt" Heckscher, a zoologist working for the Delaware Natural Heritage Program, caught an unusual clubtail in a survey of the Nanticoke watershed. Nothing like it was known from the Coastal Plain or anywhere nearby (44).

As with any unusual discovery, the specimen was examined in great detail. It did not key out cleanly, and locals briefly entertained the fantasy that is was an undescribed species before concluding that it was already known, the Midland Clubtail. But questions remained. Where did it come from? Was it a stray, or was there a population nearby? Two years later, on a canoe expedition exploring Broad Creek near its confluence with the Nanticoke River, Kitt and I discovered a thriving population that was practically inaccessible except by boat. It was mid-May, a time of year when few other species are around, certainly not the hoards of skimmer dragonflies [48-84] that populate the area in the mid-summer.

We observed only male Midland Clubtails. From their sunny perches on logs and leaves near shore, they flew out to challenge other males that happened by. At one moment none would be visible, and suddenly three or four would appear, chasing each other back and forth. They were difficult to net on the wing from a canoe; however, a different approach proved successful and much easier. As with other clubtails, they like to perch on flat surfaces near the water. An open net laid horizontally across the canoe gunnels proved irresistible. They literally landed on the net and were captured with a quick upswing.

The Midland Clubtail is a deserving member of the clubtail group, having one of the most prominently clubbed tails of any of the species known from the Delmarva Peninsula. Furthermore, the club is brightly marked with noticeable two yellow spots on each side.

While the species is widespread in the Midwest, we know very little about it along the East Coast. It is known from the Potomac River near Washington, DC, and it may be expected on the lower Susquehanna River. I had a glimpse of what might have been this species on the tidal Christina River at the outlet of

STEVE COLLINS

The Midland Clubtail, only known so far at one freshwater tidal tributary of the Nanticoke River, may be found to occur more widely if similar habitats are explored in May or June. This population differs from most in that it has a yellow spot on the top of segment nine.

Churchman's Marsh. One wonders whether it is much more widespread in the eastern United States, but rarely seen because its preferred habitat is inaccessible and its flight season is early. Explorers of other tidal freshwater habitats on Delmarva in May and June might discover additional populations.

20 DRAGONFLIES IN STONE
Gomphus lividus (Ashy Clubtail)

The late Frank Carpenter, insect paleontologist at Harvard University, had a paperweight in his office that I coveted. It was a piece of shale about eight inches long containing the fossilized impression of a single dragonfly wing almost as long. He had found it in a corn field in Elmo, Kansas, just after carefully negotiating a barbed wire fence on his way to a shale quarry where many insect fossils had been found (8). Because the fossil wing was loose on the ground and could not be clearly associated with a specific rock layer, its scientific value was diminished and that qualified it for paperweight status rather than a numbered specimen in a drawer.

That wing came from a dragonfly larger by far than any species that exists in the world today, but it was considerably smaller than *Meganeura monyi*, the largest insect known—a fossilized Permian dragonfly found in France with a wingspan of over two feet. Scientists have speculated that dragonflies of that size could exist then because birds which would out-compete them had yet to appear and oxygen concentrations in the atmosphere were higher than today, permitting the increased metabolism needed to sustain a larger body.

I have no idea how abundant dragonflies were 300 million years ago, but considering the special circumstances required to make a fossil, I wonder whether any of last year's dragonflies avoided being eaten, did not decompose, and are destined for fossilization. Only once have I seen the first step in the process. That was when I discovered an Ashy Clubtail stuck upside down in fine mud on a stream bank in Maryland. I imagined that somehow in the next step it could be gently covered over and protected with a layer of silt and would remain buried for millions of years under accumulating sediment. Its impression would be left in the silt and mud that over time would transform slowly into shale, a sedimentary rock containing it as a fossil dragonfly. More likely this Ashy Clubtail would have been discovered by ants or a passing bird ten minutes after I photographed it. But even if it had not been found and eaten or decomposed, the landscape on Delmarva is eroding rather than accumulating and sooner or later a flood undoubtedly would wash it away and destroy it.

Ashy Clubtails are no more likely to be fossilized than other species other than that they are fairly common on the Delmarva in May and June near streams. They are mostly brown with markings that lack the high contrast definition of many other clubtails. Typically they perch flat on sunlit rocks and logs along the shore or sticking out of the water. As with many clubtails, they often feed in fields or other open places away from water for several days after they emerge and before returning to their breeding sites. When disturbed in these feeding areas, they often fly in a series of up-and-down "roller coaster" motions, changing direction as they go, before perching again. While engaging in these maneuvers, they are difficult to follow and may elude potential predators, including people.

Somehow the Ashy Clubtail (right) got stuck upside down on wet mud and could not escape. Such rare occurrences millions of years ago initiated the process that produced the dragonfly fossils found today in Kansas and elsewhere. Ashy Clubtails normally perch on logs, rocks, and bare banks along streams. They lack yellow on the sides near the end of the abdomen and are slightly larger than the Lancet Clubtail [18].

21 A THREATENED SPECIES
Gomphus rogersi (Sable Clubtail)

Before 2007, Blackbird Creek in Delaware's Blackbird State Forest was the only known place on the Delmarva Peninsula where the Sable Clubtail occurred. In 2007, a new site was discovered not far away on Unicorn Creek near Millington, Maryland, and the following year another appeared in Sussex County Delaware. These are among the very few suitable habitats on the Coastal Plain for this black and pale gray-green species that is normally associated with sandy-bottomed spring-fed streams in the Appalachians. The Sable Clubtail flies in May and June, typically perching on streamside vegetation in sunny glades or abandoned beaver meadows. Periodic visits show that the Blackbird Delaware population, though localized, has remained stable since 1980, while some significant threats to its habitat have come and gone and others loom.

In the 1980's, Gypsy moth caterpillars infested Blackbird State Forest and caused significant defoliation in the area. Those few people who knew that the Sable Clubtail lived there feared that insecticides sprayed from planes to control the caterpillars would be the demise of the population. But it survived. Then logging became an issue, and some were concerned that erosion from logging operations would cause siltation that would destroy the sandy substrates the larvae prefer. Fortunately, good management practices avoided that problem.

When walking in the forest near Blackbird Creek, one feels isolated from the bustling world of super highways and shopping malls, yet the sobering reality is hard to ignore when looking down from 20,000 feet on a commercial airliner approaching Philadelphia. From that vantage point, the patches of forest are clearly seen to be surrounded by farms that can harm fragile aquatic habitats with runoff containing fertilizer and insecticides. Nevertheless, the farms have been there for a long time and the Sable Clubtail has survived.

More threatening to the Sable Clubtail is the explosion of housing developments in southern New Castle County that are devouring farmland. Houses become the last crop as farmers cash in on their land. Urbanization with its runoff containing homeowners' insecticides, fertilizer, oil, and gasoline does not treat watersheds and small streams kindly, so there is reason to worry that the Blackbird Creek population of the Sable Clubtail will be lost without vigilance and cooperation. Education about habitat conservation is key to the long-term survival of many threatened habitats and their associated plants and animals.

The Nature Conservancy (55), working with local landowners and other stakeholders, has developed a comprehensive plan for the Blackbird-Millington Corridor. Compared to nearby highly agricultural areas to the north and south, this is a relatively forested swath of land that cuts east-west across Delaware from salt marshes on Delaware Bay into Maryland. Hopefully, this plan will provide the guidelines for preserving not only two of the three now known Delmarva sites for the Sable Clubtail, but also the habitats of a number of other species that are unique or distinctive for this area. Managing wetlands is tricky business. In particular, we don't know the critical survival factors needed to sustain rare species with highly specialized habitat requirements. While striving to identify these factors, we need to err on the side of conservation.

In January 2005, snow on farm fields high-lighted the isolation of the forest tract where Blackbird Creek flows diagonally through Blackbird State Forest in the lower third of the aerial photograph above. The male Sable Clubtail (right) perched on a skunk cabbage leaf was photographed there in early June. It belongs to one of only three populations known on the Delmarva Peninsula.

22 THE *TYRANNOSAURUS REX* OF THE DRAGONFLY WORLD

Hagenius brevistylus (Dragonhunter)

On first sight, everyone finds the Dragonhunter impressive. It has a take-your-breath-away appearance. This, our largest clubtail, is the *Tyrannosaurus rex* of the dragonfly world. Its massive thorax, colored black with yellow stripes, seems to dwarf a relatively small head with bright green eyes that conceal powerful jaws below. Its long spiny legs can capture large prey in flight. It prefers to eat large insects like butterflies and other dragonflies, hence its name. As far as I know, it does not eat others of its own species; so it is not any more cannibalistic than humans who eat the flesh of other mammals.

Typically, Dragonhunters find a dead stick perch with good visibility along a stream or river bank and wait for food to fly by. I have seen them catch and eat dragonflies that are nearly their size, like the Illinois River Cruiser [34]. In eastern North America from southern Texas to Canada, there is nothing else like it. As a top predator in the dragonfly food chain, it is not common on the Delmarva Peninsula; but it is known from most of the larger streams including the Pocomoke River in the south to the Big Elk and White Clay Creeks in the north.

JIM WHITE

Dragonhunters, like this male, find prominent perches along stream banks and wait for their next meal to fly by.

The larva of the Dragonhunter is impressive as well. Full grown, its abdomen has the general size and shape of a half dollar. Like the adults, it also has long legs, but the rest of its sprawling flat body needs considerable remolding to achieve the adult form. Its resemblance to a leaf in the stream leaf litter where it lives appears to provide camouflage that protects it from being eaten by hungry fish. Despite its size and striking appearance in isolation, it often evades the human eye when found with organic stream debris. Emergence starts in late May, and adults have been found into early September.

With respect to their treatment of females (and other males), male Dragonhunters have a caveman reputation. The clamp-like appendages at the end of the abdomen in male clubtail dragonflies often are species-specific and fit into complementary depressions at the back of the female's head. Examination of the heads of female Dragonhunters reveal cracks and dents to their eyes and exoskeleton made by the male appendages as if those complementary depressions were made by force rather than by an evolutionary process. Sometimes males also show similar head damage (35).

The abdomen of a female Common Sanddragon [24] disappears in the mouth of a Dragonhunter (left). The dark brown, leaf-like nymph (larva) of the Dragonhunter, seen here against a white sand background, is easily missed when found among the leaf-litter in streams.

23 *UNIDENTIFIED SNAKETAIL

Ophiogomphus incurvatus incurvatus (Appalachian Snaketail)

The logistics of downstream canoeing normally involve shuttling people and cars between boat launch and landing sites; however, it can be a solo venture with effort and proper circumstances. One gorgeous early June day in 1973, I decided it was time to survey the dragonflies on White Clay Creek north of Newark, Delaware. The downstream canoe trip was notable for being the only time I have seen snaketails on the Delmarva Peninsula. Unfortunately, I was unable to catch and identify the few I saw. Collecting or photographing dragonflies from a canoe, even when it is a cooperative venture with net or camera and paddle, is not recommended.

None of the publications on dragonflies include the Delmarva Peninsula in the distribution of snaketails. However, because snaketails as a group are easy to recognize, I feel comfortable including an additional genus, if not a particular species, in this book. When a species designation is in doubt, it is listed as "sp." Based on habitat (Piedmont stream) and geography (Middle Atlantic), the bright

Bob Moul photographed his male Appalachian Snaketail in nearby York County, Pennsylvania. If it occurs on the Delmarva Peninsula, it will be found on a clean flowing Piedmont stream.

green *Ophiogomphus* sp. individuals I saw flying low over the riffles were the Applalachian Snaketail, a species I have never seen close up. This species is closely related to other Snaketails and was described only recently (*19*), after my sightings. I sometimes wonder whether I would have recognized this dragonfly as a new species had I been able to catch it. Subsequently, I may have seen another not far away near Big Elk Creek in northeastern Maryland. Others have found the Appalachian Snaketail on the Pautuxent River in Maryland and in southeastern York County, Pennsylvania, on tributaries of the Susquehanna River.

There are quite a few snaketail species in the eastern United States. Males have bright green thoraxes and dark-colored abdomens with yellow snake-like markings. Males typically perch on rocks in the middle of riffles or on stream-side vegetation. Females have colors and markings similar to males but usually stay in vegetation somewhat back from the stream except when ovipositing. Both males and females are often wary when approached. The flight season for the Appalachian Snaketail in nearby states starts in early May and is over by mid June.

The only other snaketail that might be encountered on the Delmarva Peninsula is the Rusty Snaketail (*Ophiogomphus rupinsulensis*). It is bigger and prefers larger streams and rivers than the Appalachian Snaketail. It should be expected in Cecil County near the Susquehanna River where it might be found feeding in grassy fields and on hillsides.

---

*The Appalachian Snaketail is one of three species included in this book that have not yet been confirmed to occur on the Delmarva Peninsula.

24 LIVING LIKE A MOLE
Progomphus obscurus (Common Sanddragon)

Most insect field guides deal with adults rather than larvae. Immature dragonflies and damselflies, or larvae, live under water, pretty much out of sight. In contrast to the adults, larvae are present year round, more abundant than adults, and localized in defined habitats. In most cases the larvae can be identified to species. Anyone with a little effort and tolerance for getting wet in cold weather can extend their study of dragonflies to all times of the year, although I don't recommend breaking ice in this pursuit. Full-grown larvae collected in the late winter and early spring survive well in an aerated aquarium, where they can be reared to adults with little trouble, so long as they don't eat each other and they have a place to crawl out of the water to emerge.

The Common Sanddragon has one of the most interesting larvae. Appropriately, the larvae live in sandy-bottomed streams where they emerge in late May on sandbars barely above the water line. The compact, inch-long, bullet-shaped larvae of the Common Sanddragon burrow quickly in loose sand. Their legs, like those of moles and other fossorial animals, are adapted for digging. The curved tips of the front and middle pairs of legs look a bit like scoops ready to sweep sand out of the way.

I have found Common Sanddragon larvae on the Choptank and Nanticoke rivers and undoubtedly they also could be found on the Pocomoke River and Blackbird Creek where adults are known. As with many clubtails, the larvae are usually much easier to find than the elusive adults. Researchers sampling in some streams in South Carolina have used seines to collect Sanddragon larvae after electroshocking, rather than by dredging (*109*).

The adult Common Sanddragon is our only clubtail with wing markings, a small dark spot at the wing bases. It also differs from other clubtails in this area by its brown color with pale yellow or whitish markings and no green markings. The appendages at the end of its abdomen are white. Its legs are small. Males perch on sandbars and bark-covered logs in the water where they are well camouflaged and quite wary. The mature adults fly during June and July.

RICHARD ORR

HAL WHITE

Common Sanddragon larvae (left) are well adapted for burrowing in the sand. Note the scoop-shaped ends of the first and second pair of legs. Adults emerge from their larval skin on the banks of sandy streams almost at the water's edge. The adult male Common Sanddragon (above) perches on sandbars and logs near the water.

25 UP CLOSE AND PERSONAL
Stylogomphus albistylus (Eastern Least Clubtail)

Few people really get to know in any depth the life and behavior of another species other than their pets and farm animals. How about getting to know one species of insect really well? That is what Michael Blust, now Professor of Biology at Green Mountain College in Vermont, did as a graduate student at the University of Delaware. In the late 1970's, he spent a lot of time wading in the headwaters of White Clay Creek in nearby Pennsylvania studying the life history of the Eastern Least Clubtail. He sampled larval populations at different places in several streams, recorded the water temperature, counted the number of nymphs (larvae), measured the nymphs, trapped adults when they emerged in early June, and determined when they returned to the streams to reproduce.

Among the discoveries Mike Blust made was that larvae of the Eastern Least Clubtail coexist in three discrete sizes, which indicate at least a three-year life cycle (11). That may be unusually long for dragonflies in our area where we assume annual life cycles, but then again, who really knows? Some species of skimmer dragonflies can complete a life cycle in a temporary pond in 60 days [73]. What it takes is someone with enough curiosity to dedicate time and energy to finding out about the life cycle of another species. This would be a wonderful science project for a high-school student having a healthy pond or stream nearby. And there are enough species to keep students young and old busy for years.

Blust found that life in the water for the Eastern Least Clubtail nymph is dangerous. Annual mortality is about 90% and rises to almost 100% in the coldest spring runs. There the water temperature in the summer rarely rises above 15°C, and predatory Spiketail nymphs are common [30]. Survival also is low in the larger streams. The so-called second-order streams formed by tiny spring runs (first-order streams) where nymph densities sometimes exceed 50 per square meter are optimal for survival. The hardy nymphs of the Eastern Least Clubtail that survived three or four years have a synchronous emergence in early June and leave the stream to feed and mature as adults before returning to the stream to breed about two weeks later.

The Piedmont woodland streams that abound in northern New Castle and Cecil counties are the preferred habitat for the Eastern Least Clubtail. Adults measure less than an inch and a half long, making this the daintiest clubtail on Delmarva. The body is mostly black with narrow, light-colored rings around each abdominal segment. The tip of the abdominal appendage in both sexes is conspicuously white. While I usually find wary males perched on rocks and vegetation in and near the water, they probably spend more time in the trees. Despite the abundance of nymphs in local streams, I might go for several years without seeing an adult. Here, as in the case of other clubtails, looking for the nymphs in the sand and gravel of stream beds is often the best way to determine their presence. I recognize the small, flat, brown nymph of the Eastern Least Clubtail by its whitish leg joints.

BOB MOUL

The Eastern Least Clubtail (above), though fairly common on Piedmont streams, often perches on vegetation out of view. This photograph was taken in southern Pennsylvania. The three sizes of Eastern Least Clubtail larvae (right) found in early April by Michael Blust show that the life cycle normally takes three years.

MICHAEL BLUST

26 MUD AND MOSQUITO REWARDS

Stylurus laurae (Laura's Clubtail)

When I began thinking about this book, I wanted to include all of the species of Odonata that were known or likely to occur on the Delmarva Peninsula. With feedback from Steve Roble in Virginia and Richard Orr in Maryland, I compiled two lists—one of the known and another of possible species. Naturally, I found the list of species that had not yet turned up to be more interesting because it included many species I had never seen alive. This list has provided the incentive to explore less accessible habitats with the goal of moving as many species as I could from the list of possible species to the list of species known from Delmarva. The following describes the pursuit and discovery of Laura's Clubtail.

From its origin below the dam at Mud Mill Pond at Choptank Mills in Kent County, Delaware, the Choptank River flows about two miles in Delaware before heading south-southwest into Maryland and on to the Chesapeake Bay.

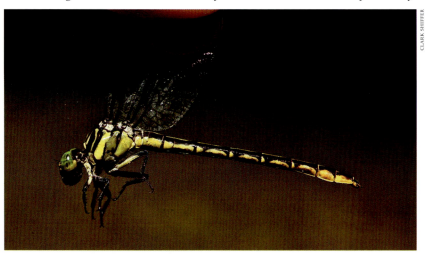

CLARK SHIFFER

HAL WHITE

Stylurus laurae Williamson, 1932   exuviae
Choptank River at south of Choptank Mills,
Kent County, **DELAWARE**
39°03.06'N, 75°44.03'W, elev. ~25 feet
Marydel USGS 7.5' Quadrangle
17 June 2006   Coll. & Det. by H. B. White   Coll. No. 06-33

This exuvia (left) is the only documentation of the presence of Laura's Clubtail on the Delmarva Peninsula. Identification is based on the shape of the hooks and spines on the labium, or lower jaw. The female Laura's Clubtail (above) was photographed in North Carolina.

North of the Delaware Route 211 bridge, the Choptank flows in braided, mostly sand-bottomed channels through an often wet and muddy forested floodplain where green stands of chest-high lizard tail grow abundantly. Above the floodplain, the land has been mined for sand, and many abandoned borrow pits are now small ponds. Although this area is impossible to transverse while staying clean and dry, and at times is home to hordes of mosquitoes and deer flies, it is one of the most interesting areas for dragonflies and damselflies on Delmarva.

On June 17, 2006, while taking an amphibious course through the floodplain, I disturbed a recently emerged dragonfly from a muddy bank near the waterline. This teneral dragonfly (not fully hardened) flew weakly on its first flight and landed among the leaves of a nearby swamp maple tree. From the setting, the time of year, and its slightly larger size, I immediately thought this was an unusual clubtail. Much to my surprise, my well-aimed swing did not produce a voucher specimen [6]. The net was empty! For the next five minutes, I scanned the shoreline with binoculars until I located the nymphal skin (exuvia) from which the dragonfly had hatched. Only later at home with a microscope and technical keys was I able to identify it as Laura's Clubtail, a species previously unknown, but expected, from the Delmarva Peninsula.

The typical habitat for Laura's Clubtail is described as small, sandy streams with moderate flow. This nicely fits the Choptank River in Delaware. According to E. B. Williamson, who described the species, Laura's Clubtail prefers to perch on leaves in sunny areas over the stream (107). The normal flight season is mid to late summer. The new emergence on the Choptank River matched the earliest date previously reported.

27 TREASURES FROM STUDENT INSECT COLLECTIONS

Stylurus plagiatus (Russet-tipped Clubtail)

Once upon a time, the biology curricula of most colleges included courses in zoology, botany, ichthyology, herpetology, mycology, and entomology, among others. As academic biology became a molecular science with indoor laboratories, it parted ways with its roots in natural history and ceased offering such courses. Colleges of agriculture and natural resources now provide refugia where some of these traditional courses persist. Now, as in bygone times, introductory entomology courses still require students to make insect collections containing specified numbers of properly mounted and labeled specimens identified as to order and family.

In the fall semester, students frequently put off making collections until it is too late and specimens in the wild are few and far between. Desperation sets in, and collections grow from specimens gleaned from car radiators, window sills in abandoned buildings, porch light globes, and from those "donated" by friends

Russet-tipped Clubtail flies from August through October on Delmarva's freshwater tidal streams.

from last year's class. Knowledgeable instructors who examine these collections become detectives. They tell tales about students with limited knowledge of biogeography who claim to have collected a tropical butterfly not previously seen within 600 miles of campus. Others have discovered seventeen-year cicadas that last emerged ten years before and then not in the fall. But also there are unexpected treasures too.

One fall, Bob Lake, then an aquatic entomologist at the University of Delaware, discovered a surprising dragonfly in a student collection. The Russet-tipped Clubtail was not known from Delaware and it was collected on October 13, an extremely late date. Most clubtails fly in the spring and early summer. Furthermore, the specimen was collected in Newport, a heavily populated suburb of Wilmington. The only water nearby was the tidal portion of White Clay Creek where it flowed past a major superfund site. Other than this one specimen, nothing seemed unusual about the collection, and the student confirmed that he had caught the specimen where and when the label said.

I subsequently explored the tidal White Clay Creek and Christina River by canoe upstream from their confluence and discovered a healthy population of Russet-tipped Clubtails in August and September. Males flew low over the water and hovered frequently. Others perched on stream-side vegetation. They are fairly large and their abdominal club is rich orange-brown color. This species also occurs on the tidal Nanticoke River and probably resides on all of the larger tidal fresh water rivers of the Delmarva Peninsula. It undoubtedly escaped detection for many years because this is not a habitat the "experts" thought would be interesting. It took a novice to teach the experts a lesson.

28 STREAM POLLUTION AND DRAGONFLY POPULATIONS
Stylurus spiniceps (Arrow Clubtail)

Because adult clubtails often are elusive, early spring larval surveys sometimes give a better perspective on local species diversity and abundance than searching for adults later in the year. This makes sense because the nymphs are concentrated in an aquatic habitat while adults of different species emerge at different times, disperse from the water, and may experience heavy mortality from birds.

As it turns out, the first records of the Arrow Clubtail from the Delmarva Peninsula are from larvae collected from the White Clay Creek by Bob Lake and his students from the University of Delaware in the 1970's. Interestingly, despite specific searching, the Arrow Clubtail was not found on White Clay Creek again until September 2007. The absence of the Arrow Clubtail for several decades may reflect the general impoverished state of the macroinvertebrate fauna of White Clay Creek below Avondale, Pennsylvania.

The Stroud Water Research Center in collaboration with the Delaware Nature Society has conducted annual surveys of the White Clay watershed and documented the decrease in invertebrate biodiversity from the clean headwaters in Pennsylvania to the large stream flowing through New Castle County, Delaware (*82*) Thus, for whatever reasons—frequent floods and silting from agricultural runoff, insecticides and other toxic substances, or something else—this nationally designated Wild and Scenic River supports far fewer dragonflies and other invertebrates than one might expect from its often pristine appearance. I often wonder what this means about the City of Newark water I drink that comes partly from White Clay Creek! Hopefully, the reappearance of the Arrow Clubtail is a sign of gradual recovery.

Ten miles east of White Clay Creek flows Brandywine Creek. Forty percent of its water passes through at least one municipal water treatment plant upstream in Pennsylvania. Despite this, the Brandywine supports a healthier odonate fauna than White Clay Creek. Among the notable components of Brandywine Creek is the Arrow Clubtail.

My first encounter with Arrow Clubtails was in mid-July 1983, when I waded the shoreline of Brandywine Creek from Thompson's Bridge (Route 92) to the Pennsylvania state line and observed hundreds of exuviae perched on vertical mud banks within a few feet of the water. Exuviae are the larval shells left behind when the adults emerge. I estimated that there must have been over 6,000. On that day, I did not see a single adult. Most likely they had dispersed from the stream to feed in neighboring fields and forest to return later as mature adults. The larva of the Arrow Clubtail is distinctive due to its elongated abdomen that serves as a snorkel of sorts as it burrows through the mud and silt on the stream bed.

Although I have never seen large numbers of adults, they seem to be most common later in the day and later in the summer as they fly low over the surface where smooth water breaks into riffles. Up close, these insects have bright green eyes and an intricate patterning of pale yellow, brown, and black on the abdomen and thorax. The Arrow Clubtail also occurs on Big Elk Creek, a Piedmont stream in Maryland. However, a photograph of a male from the Virginia portion of the Delmarva Peninsula is far from any Piedmont stream, and one wonders where it came from.

Arrow Clubtail larvae emerge in July on vertical mud banks of Piedmont streams leaving behind their cast skins that are easily recognized by the elongated tip of the abdomen. The adult Arrow Clubtail and close relatives are sometimes called "hanging clubtails" because they frequently perch with their body hanging vertically rather than a horizontally as is typical of most other clubtails. The male Arrow Clubtail above is in the hanging position while the female is in perched horizontally. Note the egg mass on the underside near the tip of her abdomen.

CLUBTAILS   57

29 TWO SPECIES FROM ONE

Cordulegaster bilineata (Brown Spiketail)
Cordulegaster diastatops (Delta-spotted Spiketail)

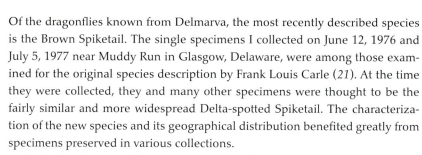

Of the dragonflies known from Delmarva, the most recently described species is the Brown Spiketail. The single specimens I collected on June 12, 1976 and July 5, 1977 near Muddy Run in Glasgow, Delaware, were among those examined for the original species description by Frank Louis Carle (*21*). At the time they were collected, they and many other specimens were thought to be the fairly similar and more widespread Delta-spotted Spiketail. The characterization of the new species and its geographical distribution benefited greatly from specimens preserved in various collections.

As is often the case when one species is split into two, not everyone agrees. Consequently, some specialists questioned whether the Brown Spiketail is a valid species and not just a geographical or morphological variant of the Delta-spotted Spiketail. Those concerns have subsided for the most part because mitochondrial DNA sequences have supported the morphological separation (*65*).

Despite numerous visits to the Muddy Run area in the three decades since my collections, the Brown Spiketail has not reappeared nor has the species turned up elsewhere on the Delmarva Peninsula. The Glasgow area, which still has a few of the appropriate seeps and tiny spring run habitats, is now bounded by a large corporate industrial park and the rapidly growing number of businesses along the US Route 40 corridor. It also lies next to a historic tract of land slated for a controversial housing development.

Although not common, the Delta-spotted Spiketail still survives in this area. Its habitat preference overlaps that of the Brown Spiketail, but the Delta-spotted Spiketail prefers larger, slow-moving streams with mud and detritus bottoms to spring seeps. The Delta-spotted Spiketail sometimes frequents the small rivulets below active beaver dams and the streams that flow through the meadows of abandoned beaver ponds. Although the patterns of yellow thoracic stripes and abdominal spots are similar in the two species, the background color of the Delta-spotted Spiketail is black, rather than brown. Both species fly from May into July and often perch on sticks along the small stream habitats that they patrol.

The Brown Spiketail (top) has markings similar to the closely related Delta-spotted Spiketail (bottom). Both are rare on Delmarva.

30 COOL SPRINGS IN THE SUMMER
Cordulegaster erronea (Tiger Spiketail)

The small tributaries that feed White Clay Creek and other large Piedmont streams in northern New Castle and Cecil counties have their sources in numerous woodland seeps. While a few of these perennial springs bubble up out of the ground, most arise in moist hillside patches with lots of decaying leaf litter and luxuriant stands of skunk cabbage. Their cold water rivulets join to form small, sand and gravel-bottomed streams—home of the Tiger Spiketail.

Unlike other spiketails, the Tiger Spiketail flies all summer. It also stays close to its preferred habitat. Because it is not common and is virtually the only dragonfly on these small streams, the streams may seem barren of dragonflies at a time of year when local ponds are swarming with them. When looking for the Tiger Spiketail, I find a comfortable place to sit with a good view of a perennial, spring-fed, woodland stream and wait for ten minutes while enjoying the shade on a hot summer day. If nothing appears within the ten minute test, I may move on, or, if the day is pleasant, I will extend the test to thirty minutes. Although I have never waited all day, at different times I have observed patrolling males from mid morning until dusk. Therefore, if Tiger Spiketails are around, the ten minute test usually is sufficient to spot one.

Occasionally, I have been lucky enough to observe a female Tiger Spiketail laying her eggs in a way that is characteristic of other spiketails as well. The female hovers with her body horizontal about eight inches above shallow water only an inch or two deep and moves up and down like a pogo stick, each time dipping her abdomen vertically down through the water and depositing an egg into the stream bottom. Little wonder that the ovipositor, which gives spiketails their name, is long and sturdy. The end of the female's abdomen is soft and can bend when thrust against bottom sediments.

More than once I have become aware of a Tiger Spiketail that has at first evaded my eye and then suddenly appeared, phantom-like, almost at my feet. That might seem hard to believe when examining the insect close up. Nearly three inches long, it has large bright green eyes and bright yellow rings around each black abdominal segment. Yet, surprisingly, this spectacularly beautiful insect is well camouflaged as it flies a few inches above the moving water, in and out of sun and shade, hovering frequently and then moving on.

I first encountered the Tiger Spiketail in similar habitats in central Pennsylvania. There I associated it with another impressive dragonfly, the Gray Petaltail [1], a large primitive dragonfly that has a habit of landing on tree trunks (and humans). The only Delmarva record of the Gray Petaltail is historical—a sighting near Octoraro Creek in Cecil County (40). I keep hoping that some day, while waiting for a Tiger Spiketail, a Gray Petaltail will mistake me for a tree trunk.

The Tiger Spiketail has striking yellow on black markings that actually provide camouflage in the flickering sunlight along small, spring-fed, woodland streams. This photo of a female shows the long straight ovipositor that distinguishes the spiketails. Though strikingly colorful, the male Tiger Spiketail (inset) often goes unnoticed as it flies in and out of shade along forested spring runs.

31 AQUARIUM PETS
Cordulegaster maculata (Twin-spotted Spiketail)

Sometimes I find it easier to locate certain dragonfly species in the winter or early spring as immature larvae than to find them in the late spring and summer as adults. The larvae of most species are adapted for particular habitats. This is especially so for stream species. Some larvae burrow in silt and mud where sediment settles in slower and deeper parts of a stream. Others prefer the sand or organic debris that collects in eddies downstream from rocks and logs in faster water. Yet others will cling to roots and sticks along the stream banks. And some will cling to rocks in the middle of rapids—all in the same stream. As adults, dragonflies are harder to catch because they fly, range more widely, and their abundance declines as predation and age take their toll. However, identifying larvae takes specialized knowledge in the absence of field guides. Rearing dragonfly larvae in an aquarium until they emerge as adults can solve that problem and provide hours of observing their underwater world in household comfort.

March and April are the best time to collect dragonfly larvae. By then any ice has melted, the days are warmer, and most of the full-grown larvae that are destined to hatch have finished feeding while the internal processes of metamorphosis proceed. There is no pupal stage for dragonflies and damselflies. While a kitchen strainer can work in a pinch, a dredge constructed of wire mesh with a one-eighth or three-sixteenth inch weave built around a sturdy triangular frame works best for catching larvae (see photo). Sand, debris, and smaller insects can pass through the mesh but the larger mature larvae are retained. The dredge I made over forty years ago out of a strong handle of hickory from a broken lacrosse stick, a bent quarter inch iron rod for a frame, and automobile hose clamps for attachment, still works pretty well. Periodically, I replace the wire mesh as it wears out.

I use the dredge like a snow shovel and dig under a layer of sand or debris. I then lift it to the water surface and slosh it back and forth so that the water can wash out the smaller particles. I pick through the remaining material looking for larvae. Those with well-developed wing pads will emerge within a month or two if put into a well-aerated aquarium with some sand on the bottom and sticks or rough surfaces on which to climb out of the water. I put a screen over the aquarium so that the larvae do not crawl away and newly hatched adults do not fly around the house. Earlier-stage larvae can be fed worms and other aquatic insects, but full-grown larvae no longer feed in the weeks before emergence.

Among the most interesting larvae from streams are those of the Twin-spotted Spiketail. They are robust. Smaller aquatic insects are dinner for this ominous and imposing predator with its large extendable lower jaw and many teeth. Because Twin-spotted Spiketails have a life cycle of more than a year, larvae of several sizes coexist.

Like the larvae, adults are large. They have bright yellow thoracic stripes and paired yellow dots on each abdominal segment. Mature adults have green eyes. Female spiketails are easy to identify because they have a particularly long ovipositor that extends well beyond the end of the abdomen. Adult males patrol long beats over small and medium-sized streams. If you miss them going one way, they usually come back a few minutes later. They fly from April through June on the Delmarva Peninsula.

The Twin-spotted Spiketail has striking green eyes and bold yellow spots on a black body. The larvae of the Twin-spotted Spiketail and other species can be collected from sandy stream beds using a homemade dredge, like the one shown, and reared to adults in an aquarium. The front edge of the dredge is about 15 inches wide.

32 PHOTOGRAPHIC DOCUMENTATION
Cordulegaster obliqua (Arrowhead Spiketail)

Standards are changing with respect to documenting the presence of uncommon dragonflies. This has to do with whether or not and to what extent photographs can substitute for a preserved voucher specimen [6]. As a result, disagreement exists among dragonfly experts, and standard protocols remain to be established. The root of the issue comes from increased interest in dragonflies and damselflies by amateur naturalists, particularly birders, who have no interest in collecting or maintaining a collection, and who bring with them the culture and ethics of the vertebrate world where collecting without state and sometimes federal permits is prohibited. With the wide availability of high quality digital cameras, more frequently photographs have become an acceptable form of documentation despite the sometimes legitimate protests of a few. In the fairly recent past, only a well-preserved and carefully documented specimen was acceptable.

Why might someone object to a high quality photograph for documentation? There are several reasons. For one, distinguishing among closely-related species often depends upon structural characters that are not visible in a photograph. On the Delmarva Peninsula, there are several examples of species groups where a photograph would not do [29, 78]. Another problem often expressed privately by experts is that photographs coming from casual observers may be highly interesting, but lack critical information about date and location. (Of course this can apply to specimens as well.) Another issue that taxonomists worry about is that future study may reveal that one species actually turns out to be two as was the case with the Delta-spotted Spiketail [29]. A specimen can be reexamined, whereas with a photograph, no specimen exists.

Thankfully these situations are not common, and more often than not, a good photograph is acceptable. Dennis Paulson, a highly respected birder and dragonfly specialist, recommends that important photographs for documentation be treated just like specimens in that they be printed with the relevant data on them and stored in a specific place in accessible repositories where they can be examined (*63*). However, this standard has yet to be implemented or followed consistently.

Three species included in this book are documented on the Delmarva Peninsula by single observations preserved with a photograph. One is the Arrowhead Spiketail, and the others are the Little Blue Dragonlet [56] and the Roseate

Skimmer [70]. While the Roseate Skimmer has a close relative in the Caribbean with which it might be confused, the Arrowhead Spiketail, with its distinctively shaped abdominal markings, cannot be confused with any other species.

The Arrowhead Spiketail is one of our largest species. It has striking yellow, arrowhead-shaped abdominal spots and yellow thoracic stripes. Although its geographical distribution includes most of the eastern United States, it prefers small slow streams that may dry up in the summer. Consequently, its distribution is also quite local, making it relatively rare. Jim Brighton photographed this species near Federalsburg, Maryland, on May 29, 2006. More exploration is needed to determine if there is an established population in the area and whether it occurs elsewhere on the Delmarva Peninsula.

JIM BRIGHTON

This photograph, taken by Jim Brighton near Federalsburg, Maryland, is the only documented record of the Arrowhead Spiketail on Delmarva. It likely will be found elsewhere when its preferred slow flowing semi-permanent streams are explored.

33 THE LAST WALK
Didymops transvera (Stream Cruiser)

A middle school student once stumped me with the following question, "Can dragonflies walk?" The student had read the curious fact that dragonflies were insects that couldn't walk. It seemed like a rather easy observation to make, but at the time I had never heard or thought about dragonflies walking. Sure enough, adult dragonflies and damselflies do not walk. Slightly altering their position on a perch hardly counts as walking. They fly from one place to another. Their legs are adapted for perching and for catching prey in flight, but not for walking. Typically, dragonfly and damselfly legs have prominent spines and hairs that project away from the leg so that they form a net or basket used to catch small flying insects. I suspect these would get in the way for walking.

In contrast to the legs of adult dragonflies and damselflies, the legs of their larvae are used for walking and are not used for catching prey. Just think, the last act of a dragonfly nymph is to crawl out of the water and find a suitable emergence site, sometimes many feet away from the water. Without the buoyancy of water, this last walk must be the hardest walk of a dragonfly's life.

Stream Cruisers have prominent legs both as nymphs and as adults. The larvae are sprawlers and have the appearance of a large-bodied, six-legged spider. In contrast to some species that hardly get out of the water, the Stream Cruiser gets the most out of its last walk. Stream Cruisers have been known to decorate the sides of boat houses 10 feet up with the shed skins they leave behind after emerging.

Look for Stream Cruisers in late April and May. They are most frequent in May along streams like the Marshyhope Creek at the Route 404 bridge west of Bridgeville, Delaware. Although their range includes all of the Northeastern United States, they are quite rare in the Piedmont part of our region. Adult males and females of this medium-large species look rather similar. They are mostly light brown with a diagonal cream-colored stripe on each side of the thorax and cream-colored abdominal markings. Males fly rather quickly and directly within a foot of the water as they patrol the shoreline of streams and ponds.

Stream Cruisers are brown with creamy white thoracic stripes and abdominal spots. They are most common on Coastal Plain streams in the spring.

34 DISTINGUISHABLE SUBSPECIES—ARE THEY THE SAME OR DIFFERENT?

Macromia illinoiensis georgina (Georgia River Cruiser)
Macromia illinoiensis illinoiensis (Illinois River Cruiser)

Our view of life on Earth and the classification of organisms constitutes a snapshot in time. From fossils we know about all sorts of extinct organisms and changes in organisms over time. From studies of DNA, we see the molecular continuity of life revealed. Related species, which have more recent common ancestors, have similar DNA—a similarity that decreases with the passage of time as accepted mutations accumulate in different lineages. Clearly, in the process of evolution, lineages split. At what point do we call these lineages *species*? The biological species definition speaks of reproductive isolation. If populations have the opportunity to interbreed in nature but do not, they are considered separate species. However, it takes a lot of study to demonstrate reproductive isolation. Those who think every organism neatly belongs to one species or another impose a typological view of life that does not always correspond in nature. It is perfectly reasonable that there should be intermediate forms. The river cruisers of the Delmarva illustrate the point.

The Illinois River Cruiser and the Georgia River Cruiser, now considered

The Georgia River Cruiser (above left) and the Illinois River Cruiser (above) are closely related. They differ in the size of the yellow spots on the thorax and abdomen. The Illinois River Cruiser occurs on Piedmont streams in the northern most part of Delmarva while the Georgia River Cruiser is found to the south on the Atlantic Coastal Plain. Both the spider-like larvae of River Cruisers (left) and the adults (above) have long legs.

subspecies (*31*), were originally described as separate species. Over the years, as the ranges of each were defined and the variation within each became apparent, some dragonfly specialists began to have doubts that they were separate species. In fact, from a structural perspective, except for a difference in the shape of male sex organs, they are practically indistinguishable. Color patterns that serve to separate often vary and can mislead; however, in this part of the country, humans and river cruisers can tell the difference. The Illinois River Cruiser frequents swiftly flowing streams of the Piedmont while the Georgia River Cruiser, with its more prominent yellow spots, is a denizen of the slow moving Coastal Plain streams. Whether the two ever interbreed here, like they apparently do elsewhere, we do not know. They behave here as allopatric species, those with adjacent distributions that do not overlap.

The name River Cruiser fits the behavior of these large dragonflies. Typically, they fly up and down the middle of a stream in straight lines with occasional course corrections. Usually, they turn around and come back shortly. Some have described their swift, direct flight as being like a locomotive. As might be expected, these handsome dragonflies are difficult to catch. Their flight season extends from late May through the summer into early September.

In an hour or two in July on White Clay Creek, Elk Creek, or Brandywine Creek, I will see Illinois River Cruisers several times. Only 10 miles away Georgia River Cruisers are known from the lower Christina River and are fairly common in the Pocomoke River drainage, where they often fly over woodland roads in the same way they fly over streams.

The larvae of river cruisers are large, sprawling, spider-like creatures. I find them fairly often when dredging for dragonfly larvae in larger streams. The exuviae, or shed larval skins they leave behind after they emerge, sometimes can be found months later perched on bridge abutments where they are protected from wind and rain.

35 A THORAX FULL OF MUSCLES
Macromia taeniolata (Royal River Cruiser)

It is easy to marvel at the aerial agility of dragonflies and not think about what is "under-the-hood" that enables their flight. The thorax contains a powerful engine that moves the large, long wings. One can think of the wings as a lever with the rigid side of the thorax as a fulcrum. The massive muscles of the thorax attach to the very short basal extension of the wing that forms the top of the thorax (*84*). The muscles have to be massive because the mechanical advantage is poor. The muscles extend from the top to bottom of the thorax and nearly fill the space. Over twenty different muscles have been named. A contraction pulls down a short distance on the wing extension at the top of the thorax, resulting in a substantial upward movement of the wing tip. Muscles attached close on the other side of the fulcrum pull the wings down. Interestingly, the front and hind wings do not move in unison. Rather, while the front pair is moving down, the rear pair is moving up. This can be seen readily in slow-motion photography.

The muscles themselves are a marvel and have been favorite objects of microscopic study because of their closely-packed regular arrays of long muscle fibers. In cross-section, they form hexagonal patterns while contractile fibers appear as parallel bands in the lengthwise view. Within the muscle cells, orderly arranged mitochondria, supplied directly with oxygen delivered by tiny trachea, generate ATP for muscle contraction. The muscles have a yellowish-brown color due to the high concentration of cytochromes, iron-containing proteins involved in this energy generating process.

The size of the thorax is closely proportional to the size of the dragonfly. So, one of our largest dragonflies, the Royal River Cruiser, also has one of the largest thoraces. Although it is called a river cruiser and does fly over rivers like the Nanticoke, I associate this species with large ponds. For example, the Royal River Cruiser is a regular summer resident of Lums Pond in New Castle County, Delaware, one of its northernmost populations in the United States. At Lums Pond, males patrol the shoreline with long beats. They fly fairly fast and directly, about three feet above the water and perhaps ten feet out from the shore. If you miss one going one way, wait a few minutes, and he will return going the other way.

The Royal River Cruiser flies through the summer. It has deep-green eyes and a large thorax with a dark metallic sheen and a prominent diagonal yellow stripe on each side. On the top of each abdominal segment, except near the end, there is a bright yellow spot.

During July and August, male Royal River Cruisers patrol the shoreline at Lums Pond in Delaware. This is perhaps the northernmost population of this species. The wings are moved by powerful muscles that fill the thorax and attach at the wing base on either side of a fulcrum. Rather than breathe air through their mouth, air enters a network of trachea in the thorax directly through holes called spiracles. One can be seen on the side of the thorax in the middle of the long yellow stripe (inset).

36 NOT DRAGONFLY EGGS
Epitheca (Tetragoneuria) costalis (Slender Baskettail)

A basic rule in biology says that a successful parasite does not kill its host. Although not always apparent unless observed close up, dragonflies and damselflies often carry extra baggage—small red or dark spherical objects attached to the thorax and abdomen. On first sight, most people think these are eggs, which seems particularly plausible when they are attached near the end of the abdomen. However, their true identity is quite different. They are the larval stages of Hydracarina, a group of aquatic mites that attach to and feed on other insects (77). True to form, they do not kill their hosts, but they undoubtedly extract a toll.

There are probably as many water mite species as there are dragonfly species, and they are quite common. However, only a few species of water mites—those that can tolerate part of their life cycle out of water—parasitize dragonflies. Mites have an interesting challenge when a dragonfly emerges. First, they are attached to the dragonfly nymph as it climbs out of the water, then they have to detach from the dragonfly's larval shell and reattach to the adult dragonfly before it takes off on its first flight. They stay on their adult hosts for only a few days. When they detach, they need go into water so that they can complete their

The abdomen of the male Slender Baskettail (left) narrows at segment three and has unmarked wings compared to the Common Baskettail. Frequently dragonflies and damselflies will have what look like small round eggs attached to their thorax or abdomen, but they are not. Above is a cluster of water mites attached to a male Common Baskettail.

life cycle. A clear advantage for an individual mite is dispersal—a free ride to some new pond or stream.

While 20% to 50% of dragonflies and damselflies may carry these parasites, some individuals accumulate more than others. I have seen over a hundred on one adult dragonfly, and up to a thousand have been reported on a single individual. Nevertheless, these dragonflies are quite capable of flying despite the fact that their abdomen may be distorted by the hitchhikers. I sometimes wonder if they can lead a normal life after the mites detach.

The Slender Baskettail undoubtedly is parasitized by water mites because no species seems to be immune. It is quite rare and sufficiently similar to the Common Baskettail [37] that experts sometimes have trouble telling them apart in collections. I have seen the Slender Baskettail only a few times on the Delmarva Peninsula, and I began to wonder whether I was missing them because of their similarity to the Common Baskettail. That concern was dispelled on a recent encounter. I noticed a dragonfly with a black abdomen and dark green eyes perched on a sumac branch along the Chesapeake and Delaware Canal. I immediately realized it was something different and didn't even consider that it might be a Common Baskettail. It was a Slender Baskettail. Clearly, behavior and appearance in the wild provide a great deal of information not revealed in a dead and dried specimen. The Slender Baskettail has a slender, darker, almost black, abdomen that tapers its whole length. A minority of females have a brownish stripe along the front of both wings. My records are in June, but its flight season in nearby states starts earlier.

37 NOT TOAD EGGS
Epitheca (Tetragoneuria) cynosura (Common Baskettail)

The primitive ancestors of dragonflies and damselflies likely had sharp ovipositors and used them to deposit their eggs within tough plant tissues. That conclusion comes from the observation that all damselflies and a few dragonflies (darners) oviposit in this way. To argue otherwise would imply that the shared trait evolved twice, once in each group. From an evolutionary perspective, it is more parsimonious to hypothesize that the trait existed in a common ancestor and was lost once in a more recent ancestor of a group of related dragonflies.

Clearly, depositing eggs within plant tissues provides physical protection and perhaps a nutritional supplement for embryonic development. Among the dragonflies that do not have sharp ovipositors, eggs are often simply washed from the ovipositor into the water by tapping the water surface. Whether eggs deposited in this way have greater mortality during embryonic development is unknown. However, some species have ways to protect their eggs in open water. Among them are the baskettails.

The female Common Baskettail extrudes her eggs into a gelatinous mass beneath the tip of her abdomen (*103*). This ball is quite obvious as she flies rapidly and erratically only about an inch over the water searching for a place to drop them. When touched to the water near a floating twig or plant, the eggs drop off and become attached to the floating material. The eggs unravel into a string as much as three inches long and then swell with water to a quarter inch in diameter so that they resemble strings of toad eggs except that they are at the water surface rather than resting on the bottom. One or more females will use the same oviposition spot so rather large masses of jelly-coated eggs can accumulate. This likely is an advantage to each female because the larger the mass, the lower the probability that all of her eggs will be eaten or parasitized by predators.

The Common Baskettail is a medium-sized dragonfly that flies from late April to early July. Males and females are dark brown with lighter orangish-brown markings. Males establish relatively small territories along the partly shaded and wooded edges of ponds and slow streams where they hover about 3-4 feet above the water. They also will forage in small swarms in clearings or along forest edges away from water.

This male Common Baskettail (top) has a small amount of black color at the base of the hind wing. In some individuals, this dark area is much more extensive reaching to the middle of the wing. In the lower photograph, strings of toad-like eggs of the Common Baskettail are draped over floating vegetation.

38 DRINKING AND DIVING
Epitheca (Epicordulia) princeps (Prince Baskettail)

Dragonflies are considered aquatic insects because they spend most of their lives under water. As larvae they are constantly bathed in water. When they metamorphose, they leave the water. Only during oviposition will females make extended contact with the water or actually submerge as adults; otherwise adult dragonflies are terrestrial.

The transition from aquatic life to terrestrial life presents a number of significant problems, one of which is dehydration. As a larva, dehydration is a problem only if a pond dries up or a stream runs dry; however, adults must conserve water. Certainly when they eat, they consume water in their prey, and when they metabolize their food for energy through respiration, they produce water. Despite their water-tight cuticle, they lose water through breathing. On hot dry days, they avoid over-heating and evaporative water loss by going into the shade or assuming the obelisk posture, which minimizes solar heating [16]. Despite these behavioral adaptations, they do need a drink once in a while.

HAL WHITE

Male Prince Baskettails fly over the open water of ponds and slow-moving rivers. They rarely perch. Ovipositing females sometime are seen flying rapidly and erratically over the water surface with the end of their abdomen sticking up at almost a right angle and holding a mass of eggs.

Usually it is on hot days when one sees a dragonfly crash into the water, fly up, and sometimes repeat the behavior once or twice before flying off to a perch in the trees. Although we don't know for sure, most dragonfly experts think that dragonflies drink in this way.

More than once while patiently trying to get close to a Prince Baskettail I have been frustrated to have it dive into the water a couple of times apparently for a drink and fly off. Male Prince Baskettails seem like perpetual fliers. They spend much of their time patrolling about two feet above open water, often well out from the shores of ponds and slow-moving stretches of rivers. Catching Prince Baskettails over water takes great patience and a little luck because they stay away from large objects like humans standing in the water, especially if the object moves into their territory. This is a species I saw many times before I held one in my hands.

Prince Baskettails are significantly larger than other baskettails. Due to differences in size and general appearance, they first were given a different generic name, *Epicordulia,* that now serves as a subgeneric name. Males and females look rather similar being mostly uniformly brown. They have dark brown patches on their wings arranged similarly to the dark patches on the wings of the Twelve-spotted Skimmer. However, there should be no confusion because Prince Baskettails lack any white wing markings of the male Twelve-spotted Skimmer [66] nor do they have any yellow markings on their thorax. In other parts of their range, the size of the wing marks varies considerably; for example, those in eastern Maine have almost clear wings and those in the southeast have large wing markings. Prince Baskettails fly from late May to early September. Their distinctive nymphs (larvae) have prominent abdominal spines and hooks on their backs.

39 CROSSING THE DELAWARE

Epitheca (Tetragoneuria) semiaquea (Mantled Baskettail)

New Jersey has the highest human population density of any state in the United States. One would think that would be bad news for dragonflies and damselflies. It probably is in the urban areas, but surprisingly, Sussex County in northern New Jersey claims 142 species of dragonflies and damselflies (3), more than any other county in the country! Furthermore, New Jersey has a region known as the Pine Barrens that includes many wetland habitats that are protected within state parks. Curiously, there are more than a couple of dragonflies and damselflies that occur regularly or even commonly in New Jersey, just across the Delaware Bay, that are unknown or extremely rare on the Delmarva Peninsula. While this may reflect insufficient survey work on this side of the Bay, more likely it is due to the variety and extent of suitable freshwater habitats in the Pine Barrens. Similar habitats are few and far between among the fields and farms of the Delmarva Coastal Plain where ditches have drained swamps so crops can be planted or dams have made ponds out of streams and marshes. A case in point is the Mantled Baskettail.

Bob Barber (2) surveyed the dragonflies and damselflies of Cumberland County, New Jersey, a county that borders the middle section of the Delaware Bay. He reported 97 species. Among them was the Mantled Baskettail, which he found at 17 of his 24 survey sites. He considered the species common. Contrast that with the one or two records ever made for the Delmarva Peninsula. For organisms that can fly and are associated with water, surely the Delaware Bay is not a formidable barrier to movement. If width is a problem, they could cross upstream on the Delaware River where George Washington and his troops did.

The scarcity of Mantled Baskettails on this side of the Delaware Bay is probably due primarily to the near absence of bogs and cedar swamps here. The factors that keep the Mantled Baskettail in New Jersey probably are the same ones that keep several other species there as well. However, I suspect that an additional factor for the few sightings in Delmarva is the early flight season of the Mantled Baskettail that is mostly over by the first week of June. Most survey work gets done during the summer, when the largest number of species are on the wing. The Mantled Baskettail can be recognized by its small size and extensive dark markings at the base of the hind wings, a characteristic that some Common Baskettails on Delmarva also have [37].

The Mantled Baskettail is widespread and fairly common in southern New Jersey where this photograph was taken. In contrast, it is rare on this side of the Delaware Bay and its presence needs to be confirmed.

40 TAXONOMIC REVISIONS AND CHANGING SCIENTIFIC NOMENCLATURE

Epitheca (Tetragoneuria) spinosa (Robust Baskettail)

I remember my high school biology teacher making a big point of the fact that we use scientific names because they do not change. He noted that, because common names varied locally and internationally, they were unreliable. For example, the English robin is quite different from the American robin, which actually is closely related to the European blackbird, but not closely related to the American blackbird. At the time of this discourse, I had an avid interest in sphinx moths and contradicted my teacher by noting that the tobacco horn-worm had three different generic names in less than 50 years—*Phlegothontius*, *Protoparce*, and *Manduca*. Fortunately, I had an understanding teacher who tolerated and welcomed challenges from students.

Other organisms, including dragonflies, are also subject to name changing. In the parlance of taxonomy, there are "lumpers" and "splitters" among taxonomists who study closely a group of related organisms. Typically such studies result in some revision that requires new scientific names. Fifty years ago, baskettails were in the genus *Tetragoneuria,* and I still think of them that way. However, the Canadian entomologist and dragonfly authority, Edmund Walker, concluded that they were insufficiently distinguished from Old World *Epitheca,* which had been described earlier and thus had priority (*90*). However, not everyone has accepted the change, and thus both *Tetragoneuria* and *Epitheca* are used—so much for avoiding confusion with scientific names.

One of the species affected by lumping was the Robust Baskettail, a species few will see because of its early, short flight season and its limited distribution. Look for it in cypress stands in late April and early May. Males establish territories in small, somewhat sheltered coves along the shore of ponds like Trap Pond in Sussex County, Delaware. The Robust Baskettail is mostly brown with gray-green eyes. The end of the male's abdominal appendage has a distinctive spine projecting upward and backward. Another noticeable characteristic is its rather hairy thorax that gives it a gray look. I found the Robust Baskettail rather common at Trap Pond in early May of 1981, but a similar search in 2007 yielded only one among many Common Baskettails [37]. Fortunately, a healthy population was discovered in April 2010 at Idylwild Wildlife Management Area in Caroline County, Maryland.

The thorax of the male Robust Baskettail is hairy and appears gray.
Look for it in April and early May in cypress glades.

41 TRIBUTE TO FRANK MORTON JONES (1869-1962)

Helocordulia selysii (Selys' Sundragon)

Frank Morton Jones, businessman, lepidopterist, benefactor, and charter member and former president (1928-1941) of the Delaware Society of Natural History, lived in Wilmington, Delaware (*51*). He divided his large collection of butterflies and moths among Yale University, the Smithsonian Institution, and the University of Delaware. He also collected some dragonflies that were donated to the Academy of Natural Sciences of Philadelphia.

If not for a parenthetical note in an article on Pennsylvania dragonflies and damselflies (*7*), we would not know that Selys' Sundragon ever occurred on Delmarva. Frank Morton Jones collected one on the Choptank River in Kent County [**26**], Delaware on April 14, 1937. The specimen, probably identified by Philip Calvert [**80, 108**], ended up in the collection of George H. Beatty, now housed in the Frost Museum at the Pennsylvania State University, where I located it in 2005.

This was the only record of Selys' Sundragon on the Delmarva Peninsula until May 4, 2006, when Delaware Nature Society naturalist Jim White and I rediscovered it on a tributary of the Nanticoke River in Sussex County, Delaware. Because its early and short flight season ends around mid-May, before most dragonfly enthusiasts expect to see much in the field, it actually may be more common and widely distributed than now known. A thorough search of sandy-bottomed streams where it might patrol in April and May will lead to more records of Selys' Sundragon. In 2010, Jim White and I found it on Marshyhope Creek south of Smithville, Maryland.

Selys' Sundragon is a medium-sized, dark brown dragonfly with small orangish spots on the sides of its abdomen. In addition, it has a brown spot at the base of each wing and small dark spots on some cross veins at the front edge of the basal portion of the wings. The eyes have a gray-green color. The male Selys' Sundragon flies a regular beat along the stream edge about a foot above the water. Those familiar with it in North Carolina claim it rarely flies when the temperature is over 70°F.

The author (top) catching a Selys'
Sundragon on May 4, 2006, in Sussex
County, Delaware—the first sighting on
Delmarva in almost 70 years. A male
Selys' Sundragon is on the right.

42 KILROY WAS HERE
Neurocordulia obsoleta (Umber Shadowdragon)

From time to time, residents of New Castle County in Delaware read about the rare sightings of a cougar roaming their neighborhoods. How is it possible that such a large mammal could exist in a highly populated area and go almost unnoticed? When I was growing up among the mountainous forests of central Pennsylvania, most people considered comparable sightings of natural Nittany Lions unreliable—the reporters must have mistaken one animal for another in their headlights while driving drunk late at night, hallucinating, or simply looking for attention.

Just because the signs say, "Kilroy was here," doesn't mean Kilroy was here. But here the reports of cougars are taken seriously. Not too long ago, signs posted in White Clay Creek State Park warned hikers to be on the alert for a cougar. Big cats leave believable evidence such as tracks in snow or mud, deer carcasses, scat, and hair from which DNA can be extracted even if the animals themselves are not seen. The authorities have no doubts with such evidence; nevertheless, some skeptics are unconvinced.

Some dragonflies are elusive like cougars. In more than four decades of pursuing dragonflies, I had seen adult Umber Shadowdragons only once. By contrast, I came across evidence of their presence at least six times, and all in the same way. Most recently, in late May 2006, I discovered about 20 exuviae, the shed larval skins, clinging to a bridge abutment on the Choptank River at Chris-

A male Umber Shadowdragon (left) has small dark markings on cross veins near the front edge of the wings. The larval skin (exuvia) of the Umber Shadowdragon (right) was left behind on a bridge abutment in Unicorn, Maryland.

tian Park in Caroline County, Maryland. This was my first inkling that the elusive dragonfly was on the Delmarva Peninsula. Fortunately, the exuviae of the Umber Shadowdragon have characteristic abdominal spines and bumps (see photograph), so they are difficult to misidentify when you know what to look for. To see adults, on the other hand, takes determination.

The sudden awareness that Umber Shadowdragons were in the area prompted me to take a short side trip to the Susquehanna River in Perryville, Cecil County, Maryland. Within a few minutes and just before a thunderstorm, I saw many Umber Shadowdragons and had one in my hands for the first time in my life. They flew fast, low, and erratically near the shore but were silhouetted by the reflection of the bright clouds on the water.

Like cougars, adult Umber Shadowdragons evade observation with behavior that has them moving about when humans are less likely to see them. To see shadowdragons, one needs to be at an appropriate river or lake at the crack of dawn or at dusk when they come to the water. Umber Shadowdragons are brown and have a brown spot at the base of their wings and a series of small brown spots near the leading edge of the front wings.

Undoubtedly, Umber Shallowdragons occur at other locations on Maryland's Eastern Shore and in Delaware. Evidence of their presence will likely come from careful searches for exuviae clinging to bridge abutments. The graffiti often found there may say, "Kilroy was here". The exuviae may say, "Umber Shadowdragons were here." Don't bother looking for Kilroy.

43 THE SOUND OF TWILIGHT SHADOWS

Neurocordulia yamaskanensis (Stygian Shadowdragon)

Imagine trying to catch a dragonfly in near darkness. It sometimes requires more than the sense of sight. The challenge happens in the pursuit of shadowdragons whose brief flight period occurs at dusk after the sun has set or in the morning light before the sun has risen. Shadowdragons fly low and fast a few inches above the water surface and make abrupt changes in direction. The usual strategy for catching one involves giving up prime time television for an evening and finding a nice place to watch a glorious sunset over open water. About fifteen minutes after sunset and until near darkness, shadowdragons, if present, will be seen as shadows or silhouettes over the western sky as reflected on the water. Ripples caused by flowing water or wind make them harder to follow.

As darkness falls, with crickets chirping and fireflies blinking on shore, one often hears a shadowdragon without seeing it (*28*) to the east, north, or south where the water surface is dark. Some claim to have caught shadowdragons by swinging a net at a whir of wings.

Normally people do not associate dragonflies with sounds, but the possible importance of sound in dragonfly behavior needs further study. Most male dragonflies other than the skimmers have a small projection on each side of their abdomen near its base called an auricle. It can touch the hind wings with each pass. One could imagine that the sound produced could be used as a form of communication in addition to sight for a crepuscular dragonfly. Interestingly, compared to other dragonflies, shadowdragons have an enlarged basal abdominal segment that also appears hollow. Perhaps this chamber resonates. Alternatively, it may expand and contract to modulate sound when the wings brush by the auricles.

In our area, I have found the Stygian Shadowdragon near the mouth of the Susquehanna River in Cecil County, Maryland, where it is significantly less common than the Umber Shadowdragon [**42**]. This riverine dragonfly may be more common upstream where the water is shallower with a stronger current. In contrast to day-flying dragonflies that often have bright patterns and color, the Stygian Shadowdragon is a fairly uniform brown color with relatively clear wings except for a basal brown spot. The similar Umber Shadowdragon has slightly smoky wings with small spots at cross veins on the front wing.

The Stygian Shadowdragon has clear wings beyond the basal spot. Near the spot at the side of the first abdominal segment is the auricle, a small round projection, that may strike against the hind wing to create sound. Note the hairy thorax.

44 THE CYPRESS SWAMP
Somatochlora filosa (Fine-lined Emerald)

I love to study maps—United States Geological Survey (USGS) topographic maps in particular and recently in combination with Google Earth views. They reveal interesting habitats, routes of access, and the lay of the land. Unlike road maps that show you how to get from one place to another in civilized space, USGS topo maps show you where you can get away from roads and people to explore mosquito-infested swamps, mucky bottom lands, and impenetrable undergrowth. Sometimes I imagine that the rewards of venturing into these places are in proportion to the difficulty endured. I certainly entertain that equation when searching for dragonflies and damselflies that have not turned up elsewhere, but should be present.

The great Cypress Swamp, also called Burnt Swamp, straddles the east-west, Delaware-Maryland border and forms the headwaters of the Pocomoke River. It represents the largest sparsely-inhabited area on the Delmarva Peninsula. For that reason, it has intrigued me and stimulated much map study. However, I have never really explored much of the swamp. I have ventured only around the edges and along a few roads that cross it. Despite being a large area of remote territory, it is not pristine wilderness. Humans have left their mark in many ways. Devastating fires, logging, and stream channelization have altered the vegetation and landscape. Nevertheless, the swamp has the feel of a place where exotic species with southern affinities await discovery. One species only recently found on Delmarva, the Coppery Emerald (*Somatochlora georgiana*) [129], should be there for the person who is willing to endure the hardships of exploration.

One species that I do know occurs there, and few other places, is the Fine-lined Emerald. I once encountered 10 to 20 of them flying back and forth over a dirt road in Gumboro, Sussex County, Delaware. While this species occurs as far north as the New Jersey Pine Barrens, I suspect that it is well-established in the Pocomoke watershed. The Fine-lined Emerald is fairly large with a dark body and a metallic cast. It has emerald green eyes and wavy, narrow, yellow to white stripes on each side of its thorax. It seems to fly later in the season than other emeralds, being well represented in late August into September, and in New Jersey once was recorded even in early November.

The male (top) and female (bottom) of the Fine-lined Emerald have a late flight season and can be seen into October.

45 KEEPING BACKYARD LISTS
Somatochlora linearis (Mocha Emerald)

Birders and other amateur naturalists like to make species lists. There are life lists, country lists, state lists, and county lists. And some people keep lists for all of the different birds they have seen in their backyard. Because dragonflies and damselflies normally live around water, a backyard list for a property without at least a frog pond seems a bit silly. Nevertheless, I keep a mental list of the approximately 25 species of dragonflies and damselflies that I have seen in my small backyard in Newark, Delaware, over the past 35 years. Since there are no ponds or streams nearby, it shows that these insects have high dispersal ability.

Perhaps the best entry on my backyard list is the Mocha Emerald because there are relatively few records for this species on the Delmarva Peninsula. Now, if someone asked me where to go to see a Mocha Emerald, my backyard would not be high on that list. On the other hand, I am not sure where I would suggest. In my life, I have encountered this species only once at its small wood-land-stream breeding site. Most often I have had chance encounters, like the one in my backyard in the late afternoon, as a Mocha Emerald and other species

The Mocha Emerald differs from other emeralds on the Delmarva Peninsula in that it has small, light-brown spots on each abdominal segment whereas other species tend to be all black.

congregated to feed on small flying insects over a field or in a clearing.

As one becomes familiar with the common species of dragonflies and damselflies that abound at ponds and streams, one's Odonata life list seems stalled. Knowing there are many more species to be found, it is time to visit different places, at different times of the day, and in different seasons. On some warm midsummer evening, bring a net along on a picnic in a state park, and hang around in the upland fields until dusk with an eye for foraging dragonflies. The Mocha Emerald is not guaranteed, but sooner or later it seems to show up. Another place to look is along roads and power line cuts near forested streams. Note that collecting insects in state parks requires a permit. (Swatting mosquitoes and horse flies does not.).

Unfortunately, identifying the Mocha Emerald on the wing is almost impossible, but one clue to look for is that mature individuals often have wings with a smoky brown coloration. In the hand, they are generally brown with characteristic emerald green eyes and no thoracic markings. The shapes of the male appendages (blunt-ended with a downward projecting point) and the female ovipositor (pointing perpendicularly down from the abdomen) are distinctive from that of other emeralds in our area. Expect to find Mocha Emeralds in July and August. The latest date I have seen one is in the third week of September.

46 EYES TO BEHOLD
Somatochlora provocans (Treetop Emerald)

As dragonflies go, most emeralds in the genus *Somatochlora* lack bright colors. Black is the dominant thoracic and abdominal color with perhaps a tinge of brown. A few, like the Treetop Emerald have yellow thoracic markings; however, their eyes, which may be reddish-brown soon after emergence, mature to a rich emerald green. Other groups of dragonflies and damselflies may have green eyes, but they are not emerald green. Something about those eyes combined with the habitat preferences and elusive behavior of emeralds gives them a unique mystique in the dragonfly world.

Few people who develop an interest in dragonflies escape the emerald mystique. While some emeralds are southern, most have a Canadian distribution and inhabit bogs, peat-stained streams, and alpine lakes—places that usually require some planning, effort, and discomfort to reach. The shared lore includes "emerald cities" and "*Somatochora* Highway" along New Hampshire and Wisconsin woodland roads, where eight or more different species of emeralds can be found. Roads and other rights-of-way cut through normally impenetrable thickets and wetlands, providing a place where emerald and human worlds sometimes overlap. There emeralds congregate and feed. They sometimes form feeding swarms composed of several species, often including darners [2-14]. The problem for human collectors is that these swarms often are out of the reach of a long-handled net.

JIM WHITE

The male Treetop Emerald has distinctively shaped appendages at the tip of its abdomen and deep emerald-colored eyes characteristic of it and many of its relatives.

The habit of flying high is shared by most emeralds, but this characteristic is captured in the name of one of them—the Treetop Emerald. Philip Calvert (*17*) described the Treetop Emerald in 1903 from specimens he collected in New Jersey. He noted that it "usually keeps at considerable distance above one's head, both when in flight and at rest." My experience almost a century later along sandy woodland roads in the New Jersey Pine Barrens mirrors those remarks.

While Treetop Emeralds may be frequent in the New Jersey Pine Barrens, they are rare on the Delmarva Peninsula. My only encounters occurred at eye-level on July 8, 1979, along a road on the north side of Trap Pond in southern Delaware and almost 30 years later in Idylwild Wildlife Management Area in Maryland on August 1, 2008.

The Treetop Emerald is distinctive in having two rather prominent yellow stripes on the sides of its thorax. The male appendages in side view have a bend in them that is slight compared to the accentuated arch in the Clamp-tipped Emerald [47]. These features may be useful for identification, but the emerald green eyes will be what commands attention.

47 GISS OR JIZZ
Somatochlora tenebrosa (Clamp-tipped Emerald)

The conversation goes as follows:

"Hey, Look. There's a female *Somatochora tenebrosa*."
"Where? I don't see anything."
"Over there in the shadows just above the water."
"That's fifty feet away. How do you know it's *Somatochora tenebrosa*, let alone a female?"
"Don't ask me. I just know. You'll see, if we catch her."

Such conversations recur in many variations, depending on the group of organisms, when experienced and novice naturalists go together in the field. Over time observant field naturalists learn to identify species based on subtle behavioral differences. For dragonflies, a slight difference in the height or directness of flight may distinguish one species from another. It could be something about the habitat or the time of day. Frequently, the expert cannot articulate the basis for judgment. It is just the "jizz", as birders say. It is that combination of sensory input that the experienced observer knows can only come from one particular species. Apparently jizz is a corruption of GISS, an acronym used by the Civil Air Patrol meaning General Image, Size, and Shape.

Among those who spend a lot of time observing dragonflies, emeralds have a special allure. Besides being notoriously difficult to identify on the wing, they are scarce and elusive. They have the frustrating habit of foraging high up, flying back and forth silhouetted against the sky, far out of reach of even a long-handled net. My field notes have innumerable entries reporting "*Somatochlora* sp. seen." Frequently that means the following: I am walking down a woodland road near wetlands and notice an emerald flying high overhead. I spend half an hour or more watching and waiting for it to swoop down close enough for a desperate swing, only to scare it off when it finally comes closer. It gets away and I don't know what got away. Because I see few emeralds each year and usually am not sure which species I am seeing, there is no "jizz" upon which I can make a confident call like the one in the conversation above.

In the hand, most emeralds live up to their name with beautiful emerald green eyes. Their bodies are generally dark brown to black with some lighter markings. Most species are easily distinguished by their abdominal appendages in males and ovipositors in females. The male Clamp-tipped Emerald is the most spectacular of all, with its large arching appendages that must nearly encompass the female's head during mating. Clamp-tipped Emeralds are on the wing from late June into September.

The male (above left) and female (above right and below left) Clamp-tipped Emeralds forage over forested roads and clearings near their small woodland stream larval habitats. The elaborate clamp-shape male appendage clasps the head of the female when mating. The female's downward projecting ovipositor is characteristic of the species. A female Clamp-tipped Emerald (above right) oviposits on a wet moss-covered rock in the middle of a small woodland spring run. The appendages at the end of her abdomen are held erect during her repeated tapping of the moss.

48 BE PREPARED
Brachymesia gravida (Four-spotted Pennant)

Be Prepared: that's the motto of the Boy Scouts. I try to be prepared when I go in the field, but sometimes circumstances make it difficult. Having explored creeks, swamps, and ponds of the Delmarva Peninsula for many years, I get complacent. While there are some species that I might see once in ten years, I don't go in the field expecting to see something I have never seen before and have no idea what it is. That did happen not so long ago.

As I was driving north along Delaware Route 9 returning from a trip that did not involve dragonflies, I decided to take a break and stop for a few minutes at the Woodland Beach Wildlife Area at Taylor's Gut east of Smyrna. I was feeling frustrated because business was interfering with my pleasure on a warm summer day when I would rather be outside with a net and camera. There are a few fresh-water ponds, an extensive salt marsh, and a metal observation tower next to the road there [124]. It was a place where I had stopped many times before.

As I surveyed the scene in my dress clothes, I soon noticed several unfamiliar dragonflies perched on the tips of phragmites stalks. I had no idea what they were! My attempts to get near were thwarted by a lack of proper footwear for a wet and muddy pursuit. Nor did I want to venture far in this tick-infested area without some DEET. Those dragonflies that happened to be accessible seemed to sense my predicament and the wary devils flew off as I tried to approach. I didn't have binoculars, net, or camera! Talk about being unprepared!

This experience necessitated a return trip the following weekend. By then I had an idea what my quarry was. I was successful; it was the northernmost record for the Four-spotted Pennant at the time. Subsequently others have found it in New Jersey and Long Island. This reasonably large member of the skimmer family prefers coastal habitats that may be brackish, as it was where I found it. Slightly beyond the middle of each wing is a dark oval patch. The stigma is white, in contrast to almost every other species in this area [61]. The population that continues to thrive in the Woodland Beach area is the only known location for the species in Delaware, although it undoubtedly occurs elsewhere at small ponds just above the high tide mark along the coast. It is known from sites farther south on the Delmarva Peninsula such as Ocean City, Maryland, and Assateague Island National Seashore. Its flight season ranges from late June to early September, but probably is somewhat longer. Be prepared to see it at any pond near the coast.

A male Four-spotted Pennant characteristically perched at the tip of a twig in the breeze, pointing into the wind like a miniature weather vane. Its white stigmas and the black spot on each wing are distinctive identification marks. Note the exuviae of a damselfly on the same stem.

49 DRAGONFLY ART
Celithemis elisa (Calico Pennant)

An interest in dragonflies sometimes leads to unusual situations. Once while visiting a nature center in Massachusetts, I struck up a conversation with a summer staff person about some pinned dragonflies on display. She didn't know where they were collected or who collected them, but she proudly went on to say that she really liked dragonflies. In fact, she had a dragonfly tattoo and wondered if I could identify it for her. I was a bit taken aback because the tattoo was not apparent. Fortunately, it was located below the neckline but above her shoulder blade. With a little help from a co-worker, I was able to examine the artwork discreetly. While most dragonfly art work is fairly stylized, this rendering was identifiable to the skimmer family.

Some dragonfly art takes considerable liberties with nature, and it isn't clear whether it is due to poor observation or artistic license. For example, many dragonfly earrings and pins have prominent antennae or an abdomen that curves sideways, in contrast to real dragonflies and damselflies which have hardly any visible antennae and can only curve their normally straight abdomens down and under their bodies. Tiffany lampshades often have several artistically-colored, bigger-than-life dragonflies peering outward. I was enchanted with the first one I saw and thought about buying it. However, because of its price, and having no place to put the lamp, I left it in the artist's shop. However, some years later, I could not resist a spectacular brass dragonfly doorknocker. It was not appropriate for my front door, so it decorates my office instead.

Some years ago, my wife was reading our local weekly newspaper and came across a legal notice announcing that a woman was changing her first name to "Dragonfly"! I did not know her at the time, but I met the young woman later and discovered she was an artist in Newark, Delaware, where we live. She, with others, has painted several murals on buildings in downtown Newark. Perhaps surprisingly, as far as I can see, none of the murals contain dragonfly images.

When it comes to naturally produced artistry in dragonflies, I often think of the Calico Pennant, a beautiful, smallish red dragonfly in the skimmer family. It has numerous reddish-brown and red markings on its wings and abdomen. Frequently, Calico Pennants perch facing the wind on the tips of reeds that emerge from shallow ponds. They also feed in nearby fields. In my experience, it prefers fishless ponds where its abundance can be quite high, and mass emergences in early June can be spectacular natural "artistic happenings". A few individuals

A male Calico Pennant displaying the red, heart-shaped markings on the top of its abdomen and the characteristic wing pattern.

persist into September and even October. Although the Calico Pennant has an appropriate name, some people think that an even better name would have been the Valentine Pennant to call attention to the little red heart-shaped marks on the top of several abdominal segments of the males. Females have similar markings, but their colors are more yellow than red. Regardless of its name, the Calico Pennant would be a good model for any artist.

50 FIGHTING OVER "COMMON NAMES"
Celithemis eponina (Halloween Pennant)

Donald Borror, the late ornithologist and also dragonfly expert, had the temerity in 1963 to suggest to a historic gathering of other dragonfly experts at Purdue University [84] that dragonflies, like birds and butterflies, should be given standardized common names (*12*). This, he claimed, would help arouse interest in the study of dragonflies and aid in teaching the recognition of groups and species. The purists responded overwhelmingly negatively. Leonora "Dolly" Gloyd [101] let out an audible gasp from the back of the room. Alluding to a diatribe by E. B. Williamson (*105*) against introduction of some common names in Needham and Heywood's *Handbook of the Dragonflies of North America* (*56*), she proclaimed that E. B. Williamson would turn over in his grave if he heard Borror's suggestion. Others noted that these could hardly be called "common names" if no one had ever heard of them. In fact, hardly any species had truly common names at the time. The names should be termed "English names" or "vernacular names." The whole idea was seen as corrupting. Nevertheless, Borror produced a list that included many names that nobody would recognize today, such as the "Eponina Spotted Skimmer."

JIM WHITE

Halloween Pennants are rather common on the Delmarva Peninsula. Note the brighter colors of the male of this mating pair of Halloween Pennants. The male uses the appendages at the tip of his abdomen to hold the head of the female. Meanwhile the female holds the male's abdomen with her legs and swings her abdomen under the male.

Despite the hostile response, years later others like Dennis Paulson and Sid Dunkle, who had not been at the Purdue meeting, saw merit in the idea. They proposed names, solicited comments and modifications, and eventually secured the approval from the membership of the Dragonfly Society of the Americas in 1996 (64) for a greatly revised set of "common names" that had about a 10% overlap with Borror's names. They replaced Borror's "Eponina Spotted Skimmer" with "Halloween Pennant," perhaps my favorite English language dragonfly name because the name is so appropriate. These lovely dragonflies sporting orange wings spotted with dark brown, normally perch on the tips of tall grasses and twigs where they face the wind. Look for Halloween Pennants in grassy fields near ponds where they act like miniature weather vanes. They are most common in July and a few can be found in early October, but not as late as Halloween.

Observers need to become aware of daily activity patterns that can be different for different species. For example, one morning in early July at about 11:00 AM I noted hundreds of Halloween Pennants filling the air over and around some fish hatchery ponds near Unicorn Lake in Queen Anne's County, Maryland. Mating pairs were everywhere. I happened to return to the pond later on the same day around 4:00 PM, and there were no Halloween Pennants there, even though the day remained sunny and warm. Perhaps I would have found them if I had looked in nearby fields where they may have been feeding. Clearly, common species may be missed because observers are out of sync with the daily patterns and look in the right places at the wrong time.

Undoubtedly, the introduction of "common names" like "Halloween Pennant" has contributed to the recent popular interest in dragonflies and damselflies. Unlike me, who had to learn the common names after knowing the Latin names, the new generation will be making interesting new observations about the behavior of even common species like the Halloween Pennant without having to struggle with the Latin names.

51 URBAN ETHICS
Celithemis fasciata (Banded Pennant)

Having taught at the University of Delaware for almost 40 years, I often am asked whether I have noticed any changes in the students. I can't really say that I have noticed a difference academically. There always seems to be a spectrum of abilities including some really bright students. But students have changed in other ways. They come from an increasingly urbanized environment and have watched a lot of television, which means they have had fewer outdoor experiences in the remaining and disappearing natural environments. Many students now feel uncomfortable taking a walk in the woods. The outdoors away from towns and cities is a foreign environment of no interest, to be avoided or feared. It is a place with winding, uneven dirt paths rather than familiar sidewalks. It is a place with grassy fields full of annoying insects rather than manicured lawns. And it is an place with unkempt bushes hiding poison ivy rather than trimmed hedges. Faced with a weekend in the outdoors, students often respond that there is nothing to do.

The number of students majoring in biology remains high because of an interest in health professions. There is relatively little interest in plants or biology not associated with the human body. Many biology courses with laboratories deal with molecular biology, and it is difficult to hire faculty likely to get tenure who can teach courses that represent biology as taught in the past. What worries me is that our urbanized society is becoming progressively less knowledgeable and less aware of the natural world around us. That trend has consequences that affect conscientious, but uninformed, decision makers ranging from parents and home owners to elected officials and judges whose immediate concerns do not include long-term environmental consequences.

How many dragonflies and damselflies mentioned in this book will still be found on the Delmarva Peninsula a century from now? If some species are gone, few will notice or care.

Consider the Banded Pennant, a handsome medium-sized black dragonfly with distinctive black wing markings. In 1974, I found a few at three different ponds in Sussex County, Delaware. While there have been one or two sightings by others, more than 30 years passed before I saw one again. In that time, many ponds have become loaded with algae due to over-fertilization from agricultural runoff. My inability to find Banded Pennants may have been due to local extinction (unlikely), the timing, frequency, and location of my searches (possibly), or a decline in abundance (probably), resulting from human activity (possibly). Is the possible disappearance of Banded Pennants at several local ponds cause for concern? In and of itself, it is probably not a big concern because this species is well known elsewhere and has a wide distribution. However, to the extent that this is symptomatic of a much larger problem, it is a concern [111].

Rachel Carson's book *Silent Spring* (*23*) sensitized people to the declining songbird populations, which she reasonably linked to persistent insecticides like DDT. Unlike songbirds, the loss of a few pretty insects probably won't affect decision-makers who have lived their lives surrounded by concrete and asphalt. Banded Pennants will have to make it on their own. Fortunately, insects lay many eggs, so recovery can occur quickly if conditions improve before their demise. I will keep looking, but I wonder if the most significant extinction is the connection to natural history we no longer teach in biology.

JIM WHITE

Although the Banded Pennant is not a common species on Delmarva, a good population exists in Caroline County, Maryland, where this photograph of a male was taken.

52 IN HONOR OF MISS MATTIE WADSWORTH (1863-1943)

Celithemis martha (Martha's Pennant)

E(dward) B(ruce) Williamson (1877-1933), president of the Well's County Bank in Bluffton, Indiana, could finance his travels far and wide in support of his avocational interests in dragonflies until the stock market crashed in 1927 (*29*). He published over 100 articles on dragonflies and described 92 species. Others immortalized him with the dragonfly genus *Williamsonia* and the species *Somatochlora williamsoni*. He in turn immortalized fellow dragonfly specialists and others by naming species after them, such as *Somatochlora calverti* and *S. hineana*. He named *Enallagma anna* for his wife, *Libellula jesseana* for his nephew, and *Stylurus ivae* and *laurae* for the wife and daughter of a friend. He also described *Celithemis martha* that he "Named for Miss Mattie Wadsworth, for nearly thirty years a careful and unselfish collector and student of Maine dragonflies, who collected many specimens of the species here named in her honor" (*104*). She and Annie Trumbull Slossen were well-known amateur entomologists in New England in an era when almost all entomologists were men (*112*).

I sometimes wonder how Martha Wadsworth felt about being so honored.

This male Martha's Pennant was photographed on Long Island, New York. It is only known from a couple of places on the Delmarva Peninsula where it is near its southern limit of distribution.

When she published, she was Miss *Mattie* Wadsworth, not Martha. Maybe she didn't like the name Martha. If Williamson had wished to honor her, why not *Celithemis wadsworthi*? After all, it was customary to derive species names from last names unless they were family members, in which case one avoided the appearance of naming a species after one's self. I also wonder what the reaction of a male entomologist would be if his first name were used to form a species name. I'm guessing it might be interpreted as an insult. I am sure Williamson truly meant to honor Mattie Wadsworth, but it shows how sensitivities have changed since 1922.

Based on Williamson's well-known dislike of common names for dragonflies (*105*), he most certainly would object to using the name Martha's Pennant for *Celithemis martha*. For him, common names were unnecessary and often inappropriate [50].

So far as is known, Martha's Pennant has been observed just a couple of times on the Delmarva Peninsula. That was from southern Delaware and nearby Maryland. These are among its southernmost known occurrences. One site was a sand borrow pit next to Route 13 in Blades, Delaware. Such habitats with clean water and emergent reedy vegetation suit this smallish, mostly black dragonfly quite well. Like other pennants, it perches at the tips of vertical vegetation. It is somewhat like the Double-ringed Pennant [53] except that it has a more extensive dark area at the base of its hind wings.

53 BIG EYES AND LITTLE ANTENNAE
Celithemis verna (Double-ringed Pennant)

Part of being a biologist is to have a good sense of structure and function so that one can interpret observations properly. For example, knowing how bone structure relates to locomotion and tooth shape relates to feeding habits permit paleontologists to deduce quite a bit about dinosaurs from their fossils. Were they browsing vegetarians or predatory carnivores? How fast could they move? How much did they weigh? And so on. What could a good biologist tell about dragonflies, based on the observation in the haiku by Chisoku?

> The face of the dragonfly
> Is practically nothing
> But eyes! (*113*)

Clearly, vision is the dominant sense of a dragonfly. There is little room for anything else on its head other than the mouth. As anyone who has experience catching dragonflies with a net knows, it is best to swing from behind because they can see in virtually every direction except directly behind. Dragonflies have better vision than humans. They see color that extends into the ultraviolet. In addition, they are sensitive to polarized light in the sky and that reflected off water surfaces.

The compound eyes of dragonflies contain thousands of facets, or ommatidia, that focus light coming from different directions. The larger facets that collect more light are darker. By looking at a dragonfly's eye and changing its orientation, there will always be a black region known as the pseudopupil. The size of that black region changes in different orientations. When it is largest, it reveals the direction in which the dragonfly sees best. For some species, this is in the plane of the body and wings. For others, it is above the plane. While the pseudopupil is dark, the color of other parts of the eye comes from pigments or light diffracted by regular structures in the eye.

A biologist would notice that in contrast to their enormous eyes, dragonflies have minuscule, almost vestigial, antennae. This suggests correctly that pheromones and other chemical signals recognized by insects with large feathery antennae, such as moths, are of little importance in communication for dragonflies and damselflies. It also suggests that the short dragonfly antennae are of little use in touch compared to the long antennae of certain crickets and katydids.

Like all other members of the skimmer family, the Double-ringed Pennant has large eyes that virtually cover its spherical head and tiny antennae. It is a relatively inconspicuous, fairly small, dark blue species with a small black area at the base of its otherwise clear hind wings. When young, both sexes have a wide and a narrow yellow ring around abdominal segments 3 and 4, respectively. Others note that the Double-ringed Pennant likes to perch on plants well out from shore and that it is particularly wary. Most of my limited encounters have been with single, or at most a few, individuals in June and July that have perched on vegetation along the shores of small, undisturbed, woodland ponds. Among dragonflies, this species is comparatively less well-known, having been first described in 1935.

The Double-ringed Pennant occurs locally and is found infrequently, so years may go by between sightings. Among pennants, its almost complete lack of wing markings is distinctive. As with other dragonflies, its head is dominated by its eyes, and its small antennae are difficult to find.

54 GREEN JACKET
Erythemis simplicicollis (Eastern Pondhawk)

As one of our most common and widespread dragonflies, the Eastern Pond-hawk would be hard to miss on a mid-summer trip to any pond or slow-moving stream. This medium-sized skimmer has clear wings and a bright green body that becomes blue-gray in males as they mature. The tip of the abdomen remains light-colored and the face remains green. Sometimes known as Green Jackets, the green color of Eastern Pondhawks provides some camouflage in grassy depressions near ponds, where they frequently perch and feed. Normally they stay low and tend to perch on the ground or on other flat objects near the ground. This is in contrast to other skimmer species that prefer stems and perches higher up. To golfers, it might be amusing to know that Green Jackets are undoubtedly common around water hazards at the Masters in Augusta, Georgia.

An examination of the Eastern Pondhawk's diet reveals a formidable preda-tor that has a preference for larger prey including large flies, butterflies, and even other dragonflies, such as other Eastern Pondhawks. One author de-scribed it as the most ferocious species in Ohio (*42*), a description that could be applied to other parts of its range that includes the United States east of the Rocky Mountains. Early emergences on Delmarva occur in mid-May from lar-vae that are also distinctively bright green. The offspring produced by the early season adults can complete their life cycle during the summer and generate a second peak of emergence in August and early September. A few adults linger on until mid-October.

Female Eastern Pondhawks often lay their eggs in wet mats of floating algae. Attendant males may be nearby where they take low perches around the edge of a pond or perch flat on the algal mats. Usually there are enough males at a pond in mid summer for an observer to see a rather distinctive ritualistic flight behavior. Two males will hover, with one a few inches above and in front of the other. They change their relative positions when the male in back flies up and over the other male and takes a position above and in front of it while the first male moves in a curve downward. The process may be repeated more than ten times, giving the impression of intersecting circular paths near the water surface.

On Delmarva, the Eastern Pondhawk is the only green member of the skim-mer family. One might be tempted to confuse it with members of the clubtail family [15-28] that are also greenish but rarely so bright green. Both Eastern

Pondhawks and clubtails have the habit of perching flat on the ground. Consequently, more than once I have become quite excited and stalked what I thought was a clubtail only to discover the common Eastern Pondhawk. Up close, it is easy to tell clubtails from Eastern Pondhawks. The large eyes of clubtails do not touch each other, while those of the Eastern Pondhawk do.

Young Eastern Pondhawks of both sexes are green, like the female above. As males mature they develop a blue-gray pruinosity (right). Both sexes feed on larger prey including other dragonflies. The male at right is eating a moth.

55 DRAGONFLIES WITH A PREFERENCE FOR SALT

Erythrodiplax berenice (Seaside Dragonlet)

Insects evolved in fresh water. We deduce that from the virtual absence of insects that live in the ocean. There are lots of salt water invertebrates, but insects are extremely rare among them. A few insects can tolerate salt water and an even smaller number can live in and prefer salt water. The Seaside Dragonlet is the only truly marine dragonfly in the world in that its larvae can live in seawater. On the Delmarva Peninsula, it and Needham's Skimmer [65] are the only salt marsh dragonfly species, although a few other species tolerate some salt. The Seaside Dragonlet is found in salt marshes from Nova Scotia to Ecuador and at a few inland salt lakes. Consequently, its preferred habitat occurs in a narrow strip a couple of thousand miles long and a few miles wide. This nearly one-dimensional geographic distribution limits the possibilities for gene flow and thus makes the Seaside Dragonlet an interesting organism to study for genetic reasons.

What happens if a favorable mutation occurs in a Seaside Dragonlet from the Delmarva? The rate at which that gene spreads is related to many factors including the dispersal of individuals carrying the gene to adjacent salt marshes. Perhaps the gene, favorable here, is not favorable in Gulf Coast salt marshes. By sampling the genetic makeup of the Seaside Dragonlet from populations over its entire range, biologists can identify subpopulations based on the genetic differences and learn about the process of speciation. They also may be able to determine whether the inland isolated populations were colonized recently and where they came from by their genetic signature.

A young female Seaside Dragonlet displaying the yellow-orange abdominal markings (left) that disappear with age in males (right). The Seaside Dragonlet is the only truly marine species of dragonfly in the world because its larvae can live in saltwater.

It is doubtful that populations of the Seaside Dragonlet show much genetic difference from one salt marsh to another on the Delmarva Peninsula. The distances are too short. However, there is genetic variation that most certainly gives rise to some of the individual differences seen. Mature male Seaside Dragonlets are mostly black with little color in contrast to females and young males that have yellow spots on the abdomen. Females also vary in the amount of diffuse wing coloration, with some having clear wings and others having a diffuse brownish colored area near midwing. Despite this variation, they are easy to identify because they are the only dragonflies of their size in the salt marshes.

In all the years of observing dragonflies on the Delmarva Peninsula, only once have I seen this species away from the salt marsh. That was a single individual that I followed for about a block as it flew low over a sidewalk on the University of Delaware campus. It was less than ten miles from a salt marsh, but that was far from home for this species.

During June and July, a good place to see Seaside Dragonlets is along Route 9 in Delaware where it crosses a salt marsh south of Flemings Landing. There the road is often underwater at high tide and the marsh comes up to the edge of the road on both sides for about half a mile. Despite their larvae's ability to live in seawater, Seaside Dragonlets have not evolved to cope with open water habitats. Thus few will be found at the beach.

56 LIFER
Erythrodiplax minuscula (Little Blue Dragonlet)

I have a perverse delight knowing that I have never seen a Little Blue Dragonlet alive. Certainly, I have looked for this southern, miniature, dark-blue skimmer at well-vegetated Coastal Plain ponds. Based on what I have heard from people who describe it as fairly common in parts of North Carolina, marshy ponds seem to be the right place to look.

At times I had wondered if the reason I hadn't seen a Little Blue Dragonlet was because it didn't occur on the Delmarva Peninsula. But that hypothesis seemed less likely when a small population was discovered at my favorite boyhood stomping ground in central Pennsylvania, well north and west of here. Then came the first ever report for New Jersey in September 2005 from Ocean County to the east. It certainly was only a matter of time before the Little Blue Dragonlet would expose itself on Delmarva. Sure enough, a member of a "SWAT" team of serious amateur naturalists combing the area for interesting birds, butterflies, and dragonflies probably saw one in July 2007 in Wicomico County, Maryland, but was unable to obtain a photograph or a voucher specimen. Consequently, the Little Blue Dragonlet still needed to be officially documented.

There are numerous stories about birders catching the next flight to far off places to add one more bird to their life list. Though the impulse to take a day off and drive south two hours to get my dragonfly lifer crossed my mind, somehow I resisted. It reminded me of those tainted stories about trophy hunters who went to Africa and shot a caged animal to return triumphantly with a head to mount over the fireplace and fantastic tales of the hunt. No, the Little Blue Dragonlet can wait. It is a lifer I want to discover myself. I want the excitement of discovery. I want it to happen as part of my normal routine. Who knows? I may never have that particular experience. Yet even after many years, the knowledge that there are still good things to anticipate makes every trip in the field an alluring adventure.

I imagine myself on a hot midsummer day wading knee deep through the emergent vegetation at the edge of a small pond when something nearby catches my eye. I wander over, and there among the reeds is a little dragonfly that I've never seen before. A lifer! The male looks a bit like a Slaty Skimmer [63] that never grew up, but pale blue in color. The female is brown with darker markings.

As fate would have it, I was with a fellow dragonfly enthusiast, Rick Cheicante, when he happened to photograph a Little Blue Dragonlet in the Idylwild Wildlife Management Area in Caroline County, Maryland. He showed the digital photo to me a few minutes later. So I was close, but only saw the picture of the first documented occurrence on the Delmarva Peninsula [32]. My lifer is still waiting.

RICK CHEICANTE

This photograph of a Little Blue Dragonlet is the first documented occurrence of this species on the Delmarva Peninsula. It was photographed by Rick Cheicante July 24, 2009, at Idylwild Wildlife Management Area, Caroline County, Maryland.

57 MAIDEN FLIGHT
Ladona deplanata (Blue Corporal)

The astronomical equinox around March 21 marks the beginning of spring on our calendar, but a walk in the woods suggests that only spring peepers and skunk cabbage agree. Even the first immigrant Common Green Darners wait a week or two before they cross the Mason and Dixon Line [5]. By mid-April, spring is in the air. The sun still penetrates to the forest floor, where carpets of spring beauty abound, and in wetter places, marsh marigolds display sunburst yellow flowers against their shiny dark green leaves. At the ponds, a few amplectic toads remain trilling amidst yards of egg strings. While Common Green Darners may have arrived a week or two earlier, this is the time to expect emergence of the first local dragonflies.

One of the earliest dragonflies to emerge on Delmarva is the Blue Corporal, a smallish skimmer whose males develop a dark blue pruinosity as they mature. Despite being known from the Atlantic Coastal Plain from Massachusetts south, this species had been unknown from Pennsylvania and northern Delaware. Thus, my discovery of it in April 2004 at Pumphouse Pond in Newark, Delaware, less than a mile from Pennsylvania and not much farther from my home, got my attention and provided the perfect justification to leave my desk for a few hours and to enjoy several sparkling spring days in subsequent years.

On the successful day in 2004, it was still rather cool at 10:00 AM, but the sun shone strongly through a deep blue sky, and it felt quite warm in places

Blue Corporals characteristically perch on a flat surfaces near the ground. The recently emerged male on the right has yet to develop the chalky blue pruinesence of the mature male on the left.

out of the breeze. As I strolled along the dirt road to Pumphouse Pond, I had no doubt spring had started. The swollen buds on the trees needed only a few more warm days to leaf out. There were several different butterflies on the wing. The birds sang loudly. And the air was fresh and fragrant. After walking about thirty yards along the shore of Pumphouse Pond, the telltale flutter of a recently emerged dragonfly on its maiden flight caught my eye. The partially hardened wings glistened like wrinkled Saran Wrap®, and the slight breeze determined the trajectory. Sure enough, Blue Corporals were emerging. In the next hour I saw about ten more, including one recently emerged next to its cast larval skin, or exuvia. Careful scanning of the shoreline with binoculars revealed a few other cast skins. All were about two inches above the water line on vertical stems. I saw no other dragonflies, but the dragonfly season had clearly begun. The Blue Corporal's season would extend until mid-June.

In addition to being small, Blue Corporals and their kin have the habit of perching flat on the ground. The dark patch at the base of the front wing is divided by a clear stripe. Young males and females have light-colored "corporal" stripes on the front of the thorax.

58 HABITAT DESTRUCTION AND SPECIES LOSS
Leucorrhinia intacta (Dot-tailed Whiteface)

When I think of the Dot-tailed Whiteface, it is always with a tinge of sadness. Those feelings well up every time I drive on Route 72 in New Castle County, Delaware, and come to the curve in the road by the athletic fields at Caravel Academy, less than a mile south of US Route 40. If I am not alone, I invariably say, "There used to be a Delmarva bay where that brick building is and another one across the road where that soccer field is." These vernal ponds that dot the Coastal Plain and usually dry up by July or August were more common in the past, but many have been ditched and drained to enable crop cultivation.

I had discovered the ponds on a USGS topographic map in 1972, shortly after moving to Delaware, and decided to check them out. The larger pond wasn't very big, perhaps 200 feet in greatest length and four feet deep in the wettest spring. I doubt it had a name. It was surrounded by cattails and full of emergent and floating vegetation with spatterdock and lily pads in the middle. It proved to be a great habitat for a variety of dragonflies and damselflies that prefer fishless ponds. There on July 1, 1972, I obtained the only known record for the Dot-tailed Whiteface on Delmarva—a single male perched characteristically on a spatterdock leaf floating on the water surface. At the time, I didn't think much of it. I had just moved to Delaware from Massachusetts, where the species is quite common, and I just assumed it was common here as well.

The male Dot-tailed Whiteface is named appropriately with twin yellow spots on the top of its abdomen and a white face on an otherwise all black body. As is typical of this species, this male is perched on a floating spatterdock leaf.

Some years later, while telling this story to a biology graduate student, I learned that he used to work in construction. Ironically, he had operated the bulldozer that drained the Delmarva bay and excavated the site for the foundation of the contractor's office building that now occupies the site. Only ghosts of the Dot-tailed Whiteface linger in the basement there. The thought that the building may have problems with water in the basement is little consolation for the loss of one more natural habitat.

I hold out hope that this small dragonfly still lives on the Delmarva Peninsula. However, habitats in New Castle and Cecil counties, where it would likely be found, continue to disappear or be altered. Nevertheless, there are many places yet to be searched. Being sure that a species has disappeared is difficult, and in this case, recolonization from the north is always a possibility, though an increasingly slimmer one with the loss of each small wetland.

The Dot-tailed Whiteface is easy to identify when seen. It has a porcelain white face contrasting with an all black body, save for two adjacent yellow spots toward the end of the abdomen.

59 *LOOKING FOR GOLD
Libellula auripennis (Golden-winged Skimmer)

My encounters with the Golden-winged Skimmer have been few and far between and not on the Delmarva Peninsula and not in their normal range. Every time I think I might have found one here, it has turned out to be one of the much more abundant Needham's Skimmer [65] that has strayed from its normal salt marsh habitat. The Golden-winged Skimmer and Needham's Skimmer look alike at first glance, and as a result the Golden-winged Skimmer has been listed, so far it appears mistakenly, as a Delmarva resident.

The reliable characteristics to distinguish them are subtle and difficult to see with binoculars or even in the hand. One difference is the color of the vein at the leading edge of each wing. In the Golden-winged Skimmer, it is uniformly golden while in Needham's Skimmer it is slightly darker and brownish for the first half. Another difference is the shape and orientation of the light-colored area on the sides of the thorax. In the Golden-winged Skimmer, it is diagonal and does not extend forward while it is more horizontal in Needham's Skimmer and extends further. There are also slight differences in leg color.

Robert Gibbs, an ichthyologist who also had an interest in dragonflies, and his wife lived for a while on Cape Cod, where both the Golden-winged Skimmer and Needham's Skimmer occur in a Coastal Plain environment. However, according to the Gibbs, the two species almost never occurred together. The Golden-winged Skimmers inhabited freshwater sandy-bottomed ponds, while Needham's Skimmers stayed close to the salt marshes (*41*).

Geographically, the Golden-winged Skimmer occurs on the Coastal Plain from Massachusetts to eastern Texas with scattered records in the Appalachian Mountains. As might be expected, Needham's Skimmer is confined to coastal marshes from southern Maine to southern Texas, except in Florida, where it occurs statewide.

Whenever I see a large, bright red-orange dragonfly at any freshwater habitat, I get excited and think "Golden-winged Skimmer"! In central Pennsylvania and high in the West Virginia mountains, my hopes have been realized. In my opinion, it certainly must occur on Delmarva, but no verified record exists despite the presence of suitable habitats. I and others will keep looking for the gold.

*The Golden-winged Skimmer is one of three species included in this book that have not yet been confirmed to occur on the Delmarva Peninsula.

CLARK SHIFFER

CLARK SHIFFER

It takes a practiced eye to distinguish the Golden-winged Skimmer (above pictures) from Needham's Skimmer [**65**]. The lighter coloration bands on the thorax of the Golden-winged Skimmer extend diagonally between the wings and legs while the band aligns horizontally with the body axis in Needham's Skimmer. Though certainly to be found on the Delmarva Peninsula, it has yet to be confirmed. The photographs come from Pennsylvania and Florida.

60 INVADERS FROM THE SOUTH
Libellula axilena (Bar-winged Skimmer)

For many years I considered the Bar-winged Skimmer to be very rare. After encountering my first two in 1973 around a shallow woodland pond in the Cypress Swamp near the Delaware-Maryland state line, I had only fleeting glimpses of them over the next twenty-five years. These large, handsome, black and gray skimmers have wing markings limited to a narrow black bar on the base and leading edge, spots at the middle and at the wing tips, and a touch of white at the base of the hind wing. They frequently select grand perches on dead tree twigs, sometimes more than 20 feet above the water with a panoramic view of the vernal ponds below. They seem more wary than other skimmers and fly off to other perches when approached, which makes them hard to catch.

As with many other skimmers, the appearance of young Bar-winged Skimmers is quite different from that of mature adults. This is particularly true for males that tend to become darker and more uniformly colored. Often male skimmers acquire a pruinosity, that grayish bloom of ripe purple plums. In the case of the Bar-winged Skimmers, the abdomens of young males and females are a stunning pattern of orange and yellow among darker colors.

Since 2000, I have encountered the Bar-winged Skimmer every year. This may be due in part to my visiting their preferred woodland pond habitats more often. However, it is likely to be due more to a severe drought in the southeastern United States in the late 1990's. Bar-winged Skimmers and other species that prefer vernal ponds and woodland swamps dispersed widely when their habitats dried up. When that happened, Bar-winged Skimmers appeared for the first time across our area and northward into the three southern New England states (*80*). Having come north, they established breeding populations that have persisted. But even though I encounter Bar-winged Skimmers more frequently now, they still are uncommon on Delmarva. Their abundance seems highest in June and then dwindles through the summer into August.

A male Bar-winged Skimmer (top) showing the wing markings. A female Bar-winged Skimmer (bottom) showing its distinctive coloration. Also note how the front legs are held close to the rear of the head and not used in perching. This is typical of many skimmers.

61 THE STIGMA
Libellula cyanea (Spangled Skimmer)

Along the front edge near the tip of each wing of dragonflies and damselflies is a pigmented spot called the stigma or more properly the pterostigma. Though normally rectangular, it can vary in dimensions from long and narrow to short and squarish. Its color is usually black. The stigma fills only one of the many cells in the wing. Because this feature is absent in many fossilized wings of dragonfly ancestors, by geological standards it appears to be a relatively new and presumably important part of the wing structure.

Hypotheses about the function of the stigma range from stiffening the wing to providing a critically positioned counter-balancing weight that suppresses spontaneous wing vibrations in flight (59). While it may serve those functions, it has assumed additional roles in species where the stigma has bright colors that likely serve in species recognition or predator distraction.

The Spangled Skimmer, a common medium-sized pond species, has a bicolored stigma, which makes it unique in the eastern United States. The proximal two-thirds (closer to the body) of the stigma in both males and females is white and the distal third (closer to the wing tip) is black. Otherwise, its wings are clear with a tinge of yellow between the stigma and the nodus (midpoint of the wing). In hovering flight, these bright white spots stand out as shimmering arcs framing the male's powdery light-blue body. Females have brownish markings with creamy light-yellow sides to their thorax.

From June through August, Spangled Skimmers populate the shorelines of most ponds. Males typically perch on stems where they can survey their territory. Because adjacent territories often overlap, Spangled Skimmers keep busy chasing off intruders. Usually they return to the same perch. Spangled Skimmers, particularly those that have emerged recently, forage in fields well away from their larval and territorial habitat.

HAL WHITE

Both male and female Spangled Skimmers display a bicolored stigma and the tinge of yellow on each wing. With age, the male's body becomes a uniform blue-gray color.

JIM WHITE

62 DERMESTIDS AND EXTENDED CARE
Libellula flavida (Yellow-sided Skimmer)

Anyone who has visited an insect collection in a museum knows that it has a characteristic odor. That odor usually comes from moth balls made of naphthalene or *para*-dichlorobenzene. Neither odor is pleasant to breathe, and both are potentially harmful. Nevertheless, entomologists learned long ago that insect collections need to be protected from other insects. Small beetles known as dermestids survive quite well on dried insects and can destroy a collection unless there

As can be seen with the mating pair of Yellow-sided Skimmers below, mature males have a chalky white appearance and wings with a yellow front margin. The females are much less conspicuous. They frequent boggy areas with wet sphagnum mats.

is frequent curatorial vigilance. Moth balls do not kill dermestids, but they repel them. I have had one serious infestation that ruined a number of specimens.

I have always felt that if you collect an insect specimen, you have the long-term responsibility for preserving it so that it has potential scientific value. Thus, I am particularly dismayed whenever I discover the telltale grit or pupal cases of dermestids in my collection. It requires prompt corrective action. My usual procedure is to put the collection in a freezer for at least a day because the beetles and their larvae cannot survive such cold temperatures. For good measure, I repeat the process in case eggs might have survived to hatch later.

Knowing how difficult it is to maintain a collection in good condition, I have been reluctant to donate rare or interesting specimens to nature centers where the staff are temporary and storage facilities are lacking. The few times I have relented, I have regretted it. One of those times was with the first Yellow-sided Skimmer I ever had seen on the Delmarva Peninsula. I had collected it as part of a survey in a natural area, and I returned the properly labeled specimen to the associated nature center. Unfortunately, within two years it and the other specimens I donated were destroyed by a combination of dermestids and mice.

Yellow-sided Skimmers are reasonably common in the boggy bottomlands along streams in the New Jersey Pine Barrens, but they are rare on the Delmarva Peninsula with only a few scattered records known. Throughout most of their eastern distribution they are typically encountered in low numbers. Some publications in the 1920's suggest that there were places where the Yellow-sided Skimmer was common nearby on the Western Shore of the Chesapeake Bay, but certainly that is not the case today. Yellow-sided Skimmers have a yellow tinge to the front part of the forewing, and the females have black wing tips and a light-colored stigma that could lead to confusion with female Spangled Skimmers [61]. Their flight season here is similar to that just across the Delaware Bay in Cumberland County, New Jersey, where they can be found from May through July (2). I was excited in 2008 to see the Yellow-sided Skimmer at a small boggy pond in Idylwild Wildlife Management Area in Caroline County, Maryland, where I collected a replacement for the one destroyed by dermestids.

63 JET PROPULSION
Libellula incesta (Slaty Skimmer)

One demonstration guaranteed to fascinate and excite school children involves dragonfly larvae (also called nymphs). It will work for almost any species, but works best for those with long tubular bodies. In a clear flat-bottomed container, such as a Petri dish, place one or two larvae and enough water to cover them. Place the container on the glass plate of an overhead projector and focus the giant six-legged silhouettes on the screen for all to see. With an eyedropper, put a drop of food coloring in the water behind a larva. The observed colored streams and swirls will reveal that water is going in and out of the end of the larva's abdomen.

This is the way dragonfly larvae breathe. Rectal gills line the inside of the abdomen. Constant circulation of water over these gills enables the larva to extract oxygen. When amplified in time of danger, this action enables a quick getaway. A light touch to a larva in a Petri dish will activate the escape response, and the larva will shoot forward several inches while a stream of water exits the rear in the opposite direction. When projected on a screen, this sudden movement elicits shrieks of delight from a young audience.

In contrast, damselfly larvae do not have jet propulsion [95]. However, they do something dragonfly larvae cannot. They swim like a fish by using the external paddle-like gills at the end of the abdomen like a fish tail and wiggle their bodies rapidly from side to side. Larvae of the spreadwing family are most spectacular in this respect. An overhead projector demonstration of this escape nicely adds to the previous demonstration and can lead to good discussion about locomotion and predator-prey relationships. In the ideal situation, the students have had the joy of getting wet and finding the subjects for the demonstrations themselves.

A sampling of larvae from a local mill pond for a demonstration will likely include those of the Slaty Skimmer. While the larva may be challenging to identify, adult males of this common large skimmer cannot be confused with other species. It is the only one with clear wings and a slaty black body. Typically it perches on pond-side vegetation on hot summer days and defends a territory while waiting for females. Female Slaty Skimmers have lighter coloration with cream-colored sides to the thorax and wide brown areas. With age, they become dark gray. Slaty Skimmers can be found from early June to early October.

The dark blue-gray male Slaty Skimmer on a pond-side perch is a common sight on Delmarva. Less frequently seen are its larvae that use jet propulsion when they need to move rapidly. The photograph to the right shows a jet of water coming out of the abdomen of a skimmer larva (not a Slaty Skimmer). The jet is made visible with a drop of food coloring.

64 COMING UP FOR AIR DIFFERENTLY
Libellula luctuosa (Widow Skimmer)

The fossil record and DNA sequence analysis indicate that over a half billion years ago we and other vertebrates shared a common ancestor with dragonflies and other insects. This ancestor was probably a wormlike creature living in the sea that was able to absorb oxygen through its skin or gills. Later, but still several hundred million years ago, two distantly-related descendants of the worm ventured onto land and in different ways confronted the challenge of delivering oxygen from the air to mitochondria where it is used to oxidize sugars and fats to form ATP, the universal molecular form of metabolic energy in cells.

We and other land vertebrates inhale air into our lungs where oxygen diffuses into the blood and binds to hemoglobin in red blood corpuscles. The red blood corpuscles circulate throughout the body and deliver oxygen to the tissues, where the oxygen then diffuses to the mitochondria. In contrast, insects have no lungs and no red blood corpuscles. Instead, they have small holes (spiracles) on the sides of their body, where air enters a branching labyrinth of smaller and smaller tubes (tracheae) that in flight muscles ultimately end next

HAL WHITE

If this were a motion picture, it would show the abdomen of this male Widow Skimmer expanding and contracting in a breathing motion that helps air enter and leave through small holes called spiracles in the side of the thorax and abdomen.

to the mitochondria. Because oxygen diffuses quickly over short distances aided by the rhythmic contractions of muscles, small insects obtain the oxygen they need quite efficiently. However, this solution to breathing air limits the size an insect can reach without becoming starved for oxygen. The largest dragonflies of the Carboniferous Period were still smaller than many birds today.

Observing dragonflies breathing is interesting. To see this best, watch any of our common large dragonflies perched on vegetation along the shore of most of our local ponds. The Widow Skimmer would be a good one to choose. When the air is still, one can see the abdomen expand and contract as it forces air in and out of the tracheae. This process is greatly accelerated during flight when the thorax contracts with each wing beat.

Widow Skimmers are easy to identify by their wing markings, which someone must have thought looked like a widow's garment. The inner third is black, the middle third is white in males, and the outer third is clear. Widow Skimmers fly from late May to early October. When recently emerged or when feeding, they often frequent fields and woodland edges. As with some other skimmers, the males and females have similar markings at first, but the females become darker, and the males become more pruinose as they mature. Also the white wing bands become much more prominent. When they fly out to challenge other dragonflies or catch small insects for food, they may consume more than fifty times more oxygen than when perched.

65 ROAD KILL
Libellula needhami (Needham's Skimmer)

One sunny morning as I headed south along Delaware Route 9, the scenic route that skirts and crosses Delaware Bay salt marshes, the large numbers of Needham's Skimmers flying over the road impressed me. It seemed like at least one of these large orange dragonflies was always in view. Several met their end on my windshield, causing me to consider how many had been killed that morning along that road. Even with the light traffic, I estimated that number at over a thousand. Now multiply that by the many other roads in the area and one appreciates the magnitude of the invisible slaughter taking place—arguably far larger than the number killed by all the dragonfly collectors in the country in a year. On the other hand, the abundance of Needham's Skimmers living in coastal salt marshes from Texas to Maine far exceeds the number being hit by cars, so the carnage has little impact on the population size. However, over time small selective pressures do make a difference. Cars are a recent environmental hazard for dragonflies, and one can wonder whether selection against low-flying individuals of some species like Needham's Skimmer will lead to the evolution of higher flying behavior or staying away from roads before the oil runs out. Reduced speed limits have already been suggested for cars and trains in the vicinity of habitats in Wisconsin for Hine's Emerald, a nationally endangered species (*85*).

My musings continued as I crossed the bridge over the tidal Blackbird Creek and proceeded down the corridor of roadside phragmites through the marsh. There the air was filled with hundreds, perhaps thousands, of Needham's Skimmers feeding in the lee of trees at the edge of the marsh. This was a photo opportunity, so I stopped and discovered as many perched as were flying. Sometimes there were five or more perched in the sun every three inches or so up phragmites stems. By chance I looked at the ground by the edge of the road and saw one and then more corpses being fed upon by ants. The number of dead dragonflies in this short stretch of road suggested that I had underestimated the impact of cars. The ants clearly benefited from the invention of automobiles.

James G. Needham, for whom Needham's Skimmer was named, was an influential entomology professor at Cornell University whose specialty was dragonflies. He was coauthor with H. B. Heywood in 1929 of *A Handbook of the Dragonflies of North America* (*56*), the first comprehensive treatise on the subject. Needham's Skimmer was first described in 1943 (*93*) by Minter J. Westfall, Jr., a student of Needham who served as coauthor with Needham in the second edition of the book in 1955 (57). Prior to 1943, Needham's Skimmer had not been distinguished from the closely related Golden-winged Skimmer [59], which should occur on the Delmarva but has not been documented. The Golden-winged Skimmer is not associated with salt marshes. Among the several details used to distinguish the species is that the leading edge (costal vein) in Needham's Skimmer does not have a uniform color but is distinctly darker in the basal half extending to the nodus, the kink in the front of each wing. Even the slight color differences seem to disappear in old specimens where examination and identification under a microscope is necessary. Needham's Skimmer flies from June to early September.

The distinctive, bright orange Needham's Skimmer is the most abundant dragonfly found in Delmarva salt marshes where nearby many are hit by passing cars (left). It rarely is found at inland ponds. At the top left is a male eating a horse fly it caught. On the right are five of the thousands of Needham's Skimmers perched out of the wind on phragmites stems at the edge of a Delaware salt marsh.

66 TURNING HEADS
Libellula pulchella (Twelve-spotted Skimmer)

One of the most disconcerting moves an insect can make is to turn its head to look at you. Somehow that ability, which dragonflies share with the praying mantis, suggests a higher cognitive level that we associate with animals like cats and owls—predators that have binocular vision. Herbivorous grasshoppers can't do it. Butterflies and beetles can't do it. However, in insects the ability to turn their heads has the interesting anatomical consequence that it is precariously attached to the rest of the body around a small pivot point. This narrow connection to the thorax must accommodate nerves and the passage of food into the gut. Those patient souls who catch dragonflies and wait a long time watching one before it comes close enough to make a desperate swing with a net sometimes remark, "Oh #!?&, I knocked its head off."

Apparently in flight, changes in the relationship between the head and thorax are closely monitored by fine sensory hairs. When a dragonfly visually locks onto its prey, it is able to adjust its flight so that it takes a direct line to an intercept point rather than a longer sweeping arc that would be generated by always flying directly toward the prey (*61*). Presumably, feedback between the head and the thorax enables this behavior. Another feature of this approach is that the dragonfly stays in the same place in the prey's visual field, which may delay its detection.

A common fairly large dragonfly that catches our attention and turns our heads is the Twelve-spotted Skimmer, confusingly also sometimes called the Ten Spot. The front and hind wings of males have distinctive alternating black and white patches. Altogether there are twelve black patches and ten white patches. Females lack the white patches. Interestingly, as discussed [75], the females look disconcertingly similar to females of the Common Whitetail, which in turn are quite different from the males of that species.

Twelve-spotted Skimmers inhabit marshy ponds and coves, where males have preferred perches—often an exposed twig near open water. They typically fly rapidly in a line for 30 feet or so, stop and hover for a moment, and then take off in another direction. In this part of their transcontinental range, adults first appear in mid May and disappear by mid October. If you look carefully at the head of a perched Twelve-spotted Skimmer with a pair of binoculars, you can see its head move to get a better look at insects and other dragonflies flying nearby.

HAL WHITE

BOB MOUL

The male Twelve-spotted Skimmer (top) has twelve black spots and ten white spots on its wings. The female (below) lacks the white spots and looks quite different.

67 TERRITORIALITY
Libellula semifasciata (Painted Skimmer)

At their breeding sites, most dragonflies are territorial. Males establish areas they defend by chasing off intruding competitors. On a warm, sunny summer day, this behavior occurs constantly at any pond with a good population of skimmers. Males find prominent perches overlooking the territory they will defend and normally return to that same perch after each time they fly out to challenge an intruder. Biologists interpret territoriality as competition for a predictable, defendable, limited resource that benefits the reproductive interests of an organism. By controlling a tract of favorable habitat, the male gains access to the females that come to that territory. Defense of territory requires energy and presumably successful males are desirable mates.

While all skimmers exhibit territorial behavior, I especially like to watch Painted Skimmers in this regard. They are not typically pond species. Rather, they prefer thickly vegetated, often boggy wetlands in wooded areas. Delmarva bays with flooded buttonbush thickets often attract Painted Skimmers. In some of their preferred wet meadows, little open water exists.

From a distance, males look like blurs of orange as they hover about three feet above the water and dart off to another place to hover or perch. Sometimes two males will face off, hovering within a few feet of each other and then tear off with one in pursuit of the other. The dominant male will then return at a more leisurely pace to his favorite perch or hovering spot.

Away from water and while feeding, most dragonflies are not territorial. Often males and females seem oblivious to each other as if there were a switch that turns off when business changes from mating to eating. One wonders what cues determine this.

Up close, Painted Skimmers are particularly attractive. Although they have dark brown spots on their wings at the tip, near the middle, and a small one at the base, the impression is of an orange dragonfly because the wing veins on the leading half of the wing are amber colored. The sides of the abdomen are predominately orange. The thorax has two diagonal stripes on each side. The front one is white and the other is yellow.

Painted Skimmers are one of the first species to appear in the spring. Fully mature males are often guarding territories before other large skimmers start emerging and before the trees are fully leafed out. The earliest and latest dates recorded on Delmarva are from April 16 to August 8. Because Painted Skimmers are sometimes seen with migrating dragonflies, it is possible that there is an influx of mature adults from the South every spring and that the over-wintering local population of nymphs emerges a little later. Someone could refute this speculation by observing recently emerged adults locally in the early spring.

From a perch overlooking its territory, a male Painted Skimmer will confront other males that fly nearby.

68 WEATHER AND POPULATION FLUCTUATIONS

Libellula vibrans (Great Blue Skimmer)

A concise definition of ecology is the study of the distribution and abundance of living things. It is an apt one because all life activities in interaction with the environment and other living things determine where a species lives and its population size at a particular time. The abundance of many insects is particularly subject to year-to-year variations in local weather with respect to temperature and precipitation. The Great Blue Skimmer seems to be such a species.

In most years Great Blue Skimmers are scarce and hard to find, but in wet years when woodland pools fill up with water in the spring and stay wet into the summer, Great Blues are common. They seem to repopulate the Delmarva Peninsula and north to central Pennsylvania and southern New England from areas to the south particularly during drought periods in the southeastern United States.

Anyone familiar with passenger jets at a commercial airport, who then sees a C5A giant transport aircraft land at Dover Air Force Base, knows what big is. That difference is somewhat the sense I had when I saw my first Great Blue Skimmer after years of seeing the more common skimmers. It is like a skimmer on anabolic steroids. Its light blue-gray color reminds me a little of the color of a metal aircraft dusted with flour to remove the shine. Up close, its turquoise-colored eyes and white face attract attention. Its wings are clear except for some darkness on the base, at the mid-wing, and the wing tips.

You won't find Great Blue Skimmers at farm ponds with other skimmers. They prefer wooded floodplain pools and oxbow ponds, where males perch on sunlit twigs with a commanding view of their territory. If other males are around and enter the territory, the resident immediately confronts the intruder, and a dazzling aerial chase ensues, more characteristic of a fighter plane than a cargo transport plane. Within a few seconds the resident typically returns to the same perch. When favorable conditions exist, one can find Great Blue Skimmers at wooded wetlands anywhere, but the habitats in the Pocomoke River watershed seem most reliable. They first appear in May and fly through the summer until their ponds dry up. In wet years they may persist until early October.

In the first few days after they emerge, male Great Blue Skimmers (lower) are spectacularly colorful. Over time they develop a uniform whitish pruinosity and have gorgeous turquoise eyes (top). They perch on dead twigs around wooded floodplain ponds, a habitat that varies from year to year depending on rainfall.

69 WHERE IS BATTING'S POND?
Nannothemis bella (Elfin Skimmer)

When I first moved to Delaware, I explored the insect collection at the University of Delaware to see what interesting species of dragonflies were known from the area. Among the specimens I saw was a female Elfin Skimmer, our smallest dragonfly at less than an inch long. The dainty males are light gray while the females mimic wasps with their yellow-ringed black abdomens. Though fairly common in certain bogs in New Jersey and New England, it is among Delmarva's rarest and most threatened dragonflies because the sphagnum bogs and fens that it prefers are also small and rare here. Having developed a fondness for bogs, I was anxious to locate its habitat, not only to see this dragonfly and its habitat, but to see what other bog–loving species might be found [**98**].

The small label on the pinned Elfin Skimmer collected by Frank Morton Jones [**41**] said, "Battings Pond, New Castle, DE, 6/16/??." The year on the handwritten date was unclear. I scoured my maps of New Castle County and the town of New Castle looking for Batting's Pond without success. I asked a few people who might know, but they didn't. Thus began the mystery of the missing Batting's Pond. Had its name been changed? Was it a beaver pond that disappeared with the beavers? Was it deliberately drained and farmed or turned into a housing development?

Not much later, I learned about sea-level fens in Sussex County and found the Elfin Skimmer common in those small habitats. Subsequently, an Elfin Skimmer showed up in an undergraduate insect collection. It proved to be from disturbed habitat in Kent County, Delaware. It was a hillside sand pit that had intercepted the water table so that there was a constant seepage. Given time, this habitat could have evolved into a sea-level fen with the nearby salt marsh. Unfortunately, the habitat was destroyed to make room for parking.

At odd moments for more than thirty years, I wondered about Batting's Pond and whether the Elfin Skimmer might still occur in New Castle County. So I went back to the Delaware collection to reexamine the specimen and its label. I hoped there might be other specimens in the collection from Batting's Pond that might provide clues. A possible break in the case came from examining old maps and photos in the University of Delaware's Morris Library. They clearly showed a pond in the town of New Castle, but no name was given. Perhaps it had been drained or filled in.

My mystery seems to have been solved by Kitt Heckscher, then a naturalist for the State of Delaware, who has looked at the butterfly collection of Frank Morton Jones and has examined his field notes. Most likely, the location is *Banning's* Pond in Banning Park in Newport, a place that Jones visited frequently. Preserved plant specimens indicate that there once was a fen or bog there. Most likely, Jones was referring to New Castle County rather than the town of New Castle. Finally, he had a habit of labeling specimens with only the month and year. Thus 6/16 would be June 1916.

I am quite sure there are no Elfin Skimmers in Newport or New Castle now. It is possible that other sites for the Elfin Skimmer exist in protected places in the rapidly suburbanizing Wilmington area.

At less than an inch, the Elfin Skimmer is the smallest dragonfly found on the Delmarva Peninsula and one of the rarest. It resides in bogs and sea level fens. Females (left) look somewhat like a wasp, while mature males (right) have a uniform, light gray pruinescence.

70 ON THE MOVE
Orthemis ferruginea (Roseate Skimmer)

It is easy to almost ignore a common species but get really excited about that same species occurring outside its normal range. Common species gain anonymity by their abundance. But, the same species outside its known range gets lots of attention and has a much greater chance of being captured, photographed, or written about. The Roseate Skimmer has received more than the usual attention recently because it seems to be expanding its range northward into territory where it has never been seen before. Because it is conspicuous and not likely to be confused with any other species in this area, sight records are believable [78]. By contrast, in the tropics the Roseate Skimmer has a couple of very close relatives with which it is often confused, and only a specimen is reliable documentation.

Before 1930, the Roseate Skimmer was a tropical dragonfly known only from the Florida Keys. Now this large dragonfly with a pink abdomen, purplish thorax, and clear wings is common along the Gulf Coast and is found in many states from the Carolinas to California. Since 1976, it has been a resident of Hawaii probably as a result of some hitchhiking larvae transported with aquarium plants. Since 2000, sightings from Richmond, Virginia; Washington, DC; and near Baltimore, Maryland, indicate that the Roseate Skimmer continues to expand its range northward. These records were in July. It was only a matter of time before it appeared on the Delmarva Peninsula.

The first and only record so far from the Delmarva Peninsula comes from the extreme southern tip at Fisherman Island, Virginia. The Roseate Skimmer shares that distinction with the Regal Darner [9]. While banding Monarch Butterflies, Randy Emmett photographed a fresh (young) Roseate Skimmer on October 15, 2005, which was the second Virginia record for the species.

The Roseate Skimmer typically inhabits ponds, ditches, and slow-water environments. It is fairly tolerant of brackish ponds and polluted waters. The fresh condition of the individual photographed by Randy raises the possibility that the species may have breeding sites nearby. Certainly, it is a species that should be expected farther north on Delmarva. Who knows? It may be only a matter of time before the Roseate Skimmer becomes common here and then people will take little notice of it.

This photo of a male Roseate Skimmer taken in 2005 by Randy Emmitt on Fisherman Island at the tip of the Delmarva Peninsula is the first known record of the species on Delmarva.

71 RANGE EXPANSION: CLIMATE CHANGE OR NEW HABITAT AVAILABILITY?

Pachydiplax longipennis (Blue Dasher)

In contrast to the previous species, which few people will ever see on the Del-marva Peninsula, the Blue Dasher is hard to miss. In fact, it is our most abundant dragonfly all summer long and can be found from mid-May to late October. In July, probably more than half of all the dragonflies present are this species. Not only is it exceptionally common here, it is abundant throughout the southeast-ern United States. While it is especially fond of ponds with emergent vegeta-tion, it shows up wherever there is water with the exception of small Piedmont streams. What makes this species so abundant? I suspect that it has something to do with the habitats that people have created.

It appears that the Blue Dasher has expanded its range northward on both the East and West Coasts. It is now common in parts of New England, where it was rare or absent early in the last century. Some dragonfly specialists attribute this to global warming. Others suspect that the proliferation of farm ponds has created favorable habitats where few existed before. Yet others claim that its abundance stems from its larvae, which tolerate the algal blooms in ponds fer-tilized by nutrient-rich run off. Whatever the reason, the Blue Dasher is thriving in the early 21st century while populations of some other species seem to be declining.

Male Blue Dashers select dead twigs and other prominent places as perches to survey their territory. They dart out to chase away other males or capture small insects. They are blue, but not bright blue. Rather, the dominant color of mature males is chalky blue gray. On close examination, their true beauty becomes apparent in the pastel yellow, green, and blue patterns on the sides of their thorax. Females have similar pastel colors on their thorax, but are darker and have two dashed yellow stripes running down the top of the abdomen and little pruinosity. On hot summer days, Blue Dashers commonly adopt an obelisk posture [16] where they point their abdomen directly at the sun. In this way, they minimize the amount of sunlight hitting them and avoid overheating.

One day while wading in a marsh at Lums Pond, New Castle County, Dela-ware, I noticed several tussocks of tall grass that were ringed by dragonfly wings floating in the water. Upon examination, each tussock had a resident, well-cam-ouflaged praying mantis patiently waiting for the next unsuspecting dragonfly to land. Because of their abundance, Blue Dashers were the predominant vic-tims (*13*).

The numerous dragonfly wings floating in the water below the praying mantis (top) showed that the male Blue Dasher being consumed was not the mantid's first such meal. In contrast to the chalky blue-gray pruinescence on the abdomen of males (bottom left), female Blue Dashers (bottom right) are darker with a double stripe on the top of the abdomen.

72 WORLD TRAVELER
Pantala flavescens (Wandering Glider)

The Worldwide Dragonfly Association published its first four volumes of the *International Journal of Odonatology* under the name *Pantala* for good reason. *Pantala flavescens*, the Wandering Glider, is the only worldwide species of dragonfly. It is known from every continent except Antarctica. Ships a thousand miles at sea will encounter Wandering Gliders.

To ecologists, the Wandering Glider is known as an "r-species", a designation that comes from the Lotka-Volterra Equation that describes population abundance and growth over time. The values of two terms, r and K, in this equation help characterize different life history strategies that organisms evolve. An r-species is opportunistic. It grows rapidly, has a short life cycle, and relies on its ability to find newly formed or ephemeral habitats. It produces many eggs that have low probability of surviving to adulthood. For example, a newly filled swimming pool or flooded construction pit may be colonized by Wandering Gliders within days and have adults emerging in about two months. These habitats often lack competitors and predators and may be loaded with food like mosquito larvae. However, these high-risk habitats frequently dry up before a life cycle is completed.

Species designated as "K-species", on the other hand, have fewer offspring, develop more slowly, and typically inhabit stable and reliable habitats. The dis-

Female and male Wandering Gliders are similar in appearance. The broad hind wings enables them to fly long distances gliding on the prevailing winds.

tinction between "r-species" and K-species" is a useful one for understanding life history strategies that evolve in populations. An important lesson to understand is that the r-K gradient is a continuum in which one species is simply "more K-like" than another.

Wandering Gliders have clear wings and yellow bodies about two-inches long. Their eyes are a dull maroon. Males and females look and behave quite similarly. Most often I encounter them over fields where they sometimes form mixed species swarms, feeding on small insects along with the Spot-wing Glider [73] and Black Saddlebags [83]. Most of our records are from July to October.

In keeping with their r-species strategy, Wandering Gliders do not hang around their natal home. On the wing, they travel on the wind looking for new places to populate. Frequently, they appear over fields and parking lots, sometimes perching on car antennas, and even ovipositing on the hoods of cars, which apparently seem like a reflecting water surface to them. A closer look at their hind wings reveals they are particularly wide. This enables them to glide, saving energy and permitting transoceanic flights.

73 RETENTION BASINS
Pantala hymenaea (Spot-winged Glider)

As suburban sprawl extends inexorably into farm lands and the amount of impervious surface grows with it, runoff from rain and melting snow quickly flows into storm drains and gets shunted into streams. Consequently, many elementary school students have experienced more than one "100-year flood", the level of flooding that hydrologists calculate has the probability of occurring once in 100 years. To lessen and slow runoff, new housing developments, shopping centers, and other construction projects typically are required by code to build retention basins. Depending on their design, these can become temporary wetlands that favor opportunistic species [72] with short life cycles and high dispersal ability such as the Spot-winged Glider.

Because these temporary wetlands are very young geologically compared to natural seasonal wetlands, they have not developed a diverse flora or accumulated a rich humus substrate. Cattails and phragmites colonize these barren habitats early. During the first week of October, I once visited a large, shallow retention basin containing less than a foot of water near a newly built elementary school in Middletown, Delaware. Few plants were growing from the yellow mud in and around the retention basin. In essence, it was a very large mud puddle that hadn't yet dried up after some heavy thunderstorms in July and August. The few plant stems around it had dragonfly exuviae on them, and in the water were numerous full-grown larvae of the Spot-winged Glider ready to hatch. In the tall grass nearby, several teneral (recently hatched) adults fluttered up when disturbed. I thought that they'd better harden up quickly and head south before the first frost. In this pond without frogs or fish and only occasional herons, the larvae had been the top predator for two months, but they would be lucky to get out alive before the puddle dried up or winter came.

When recently emerged, Spot-winged Gliders are a mosaic of subtle colors and patterns. By full maturity they have an orange brown coloration and a distinctive spot at the base of their hind wings. By contrast, the related Wandering Gliders [72] are mostly yellow and have clear wings. Both species have slender delicate legs, and both travel widely. They frequently congregate over fields and parking lots far from water. Often groups will fly together with other wandering species such as Saddlebags.

Once when I was in Davis, California, the town was full of Spot-winged Gliders that had hatched from the nearby rice fields, huge retention basins in a sense. Gliders fed in swarms of 5 to 50 or more individuals, mostly 20 to 40 feet above the ground. When perching, they roosted together. I saw about 20 individuals hanging from dead twigs in less than a cubic meter of space. All had their backs to the sun and their heads tilted forward. On the Delmarva Peninsula, Spot-winged Gliders are much less common. They may be seen from May through October, but are most common mid-summer.

The perched Spot-winged Glider (top) probably emerged within the day. This is apparent because the wings are very clear, except for the distinctive spot on the hind wing, and the body has not developed the dark orange-brown color typical of fully mature adults seen in the male shown below.

74 CAREFUL OBSERVATIONS PROVOKE QUESTIONS
Perithemis tenera (Eastern Amberwing)

Aptly named, this small dragonfly with amber wings occurs commonly at ponds and slow-flowing streams throughout the eastern United States. For Delmarva, the earliest record is from April and the latest is early October. They inhabit virtually every farm and mill pond on the Delmarva Peninsula. Throughout the summer, males fly just above the water surface near the shore and perch on twigs and debris similarly close to the water. Up close, their wings seem broad and stubby as does their light brown abdomen.

In contrast to the tropics, where there are several species of Amberwings, the Eastern Amberwing is unique in this area and easy to identify. It exhibits the sexual dimorphism so common among dragonflies and damselflies [75]. Females, rather than having uniformly amber wings, like the males, have clear wings with two irregular dark brown patches—but there is more to this story.

With any common species that is easy to identify at a distance, it is easy to ignore Amberwings in search of rarer species. However, careful observation sometimes reveals unusual features missed by thousands of casual observers. These observations expose unusual phenomena and pose interesting questions. Clark Shiffer, the person to ask about dragonflies in Pennsylvania, discovered two female Eastern Amberwings whose wings were uniformly amber as in males (*75*). Wondering whether this was a unique occurrence, he discovered a few other females lacking spots, but he could find nothing about this in the literature. Subsequently, I found such a female Eastern Amberwing at Lums Pond in northern Delaware, and others have turned up in New York.

Interestingly, the females of certain Darners [2] and Forktails [121] also have two (or more) female forms, one of which looks like the males. Thus the phenomenon of female polymorphism (multiple forms), though unusual in skimmers, is well-known and also occurs in tropical Amberwing species (*39*). However, the biological significance remains obscure. Here then is another example of how a solid observation can open a host of research questions. Are these females treated differently by males? Do these females behave differently? What percentage of female Amberwings have amber wings? What are the genetics of the phenomenon? What is going on? These are good projects for someone to pursue.

The female Eastern Amberwing (top) has clear wings with brown patches while the male (below) has uniformly amber-colored wings. Occasionally, females are found with wings having the uniform amber color of males.

75 SEXUAL DIMORPHISM AND MIMICRY
Plathemis lydia (Common Whitetail)

Mimicry, where one or more organisms evolve to look like another, occurs frequently among insects. This is seen with the Viceroy butterfly that mimics the distasteful Monarch butterfly and thereby gains protection from hungry birds. Green preying mantises look like the vegetation around them and the camouflage enables them to catch unsuspecting prey. There are many examples of camouflage and mimicry in nature and the selective advantage is usually apparent. But is the similarity between unrelated species always an example of mimicry? The female Common Whitetail represents a perplexing case.

Relatively few dragonflies are so common and widespread as to have acquired a common name before the recent move to give every species an English language name [50]. The Common Whitetail is among the select few that did have a true "common name". Males have a prominent black band that cuts across both wings and a bright white abdomen. They frequent ponds and slow moving streams that tend to have exposed mud flats, where they perch and set up territories. There they confront male intruders and engage in aggressive face-to-face displays and high-speed chases as they wait for females (*46*). They first appear in mid to late April and fly until mid-October.

Although similar in size and shape, female Common Whitetails do not have a white tail, nor do they have wing markings like the males. Their abdomens

HAL WHITE

The Common Whitetail gets its name from the appearance of the male (right). The female on the left perched next to a male looks quite different—both in color and wing markings. Interestingly, the female looks quite similar to the female Twelve-spotted Skimmer [66].

are brown, margined with elongated pale spots. Their wings have several dark patches. One would guess that they belong to different species. This is a particularly striking example of sexual dimorphism, which is common among dragonflies and damselflies. Males tend to be bright, showy, and conspicuous as they guard territories while females tend to have duller colors and be more secretive. In these characteristics, they are much like many birds.

The distinctive behavior of Common Whitetails near their breeding sites disappears away from water where they frequently forage in clearings and forest edges. There, several males and females may perch side-by-side in the sun on fallen logs or rocks without aggressive or sexual interactions. In such circumstances, the striking difference between males and females is readily apparent.

Curiously, the female Common Whitetail looks very much like the female Twelve-spotted Skimmer [66], another common skimmer, which in turn does not look like the males of its species either. Both species can be found together in the same habitats. Why should the females of two distantly related species look so similar when neither looks like the males of either species? This would appear to be an example of convergence and perhaps mimicry. However, no one has provided a good selective explanation for the similarity based on the principles of natural selection. One thing is clear: males of both species have a lot less trouble than humans in telling the females apart.

76 LIVING BENEATH THE ICE
Sympetrum ambiguum (Blue-faced Meadowhawk)

Where do dragonflies go in the winter? That is a question I often hear from schoolchildren. Although following an unusually warm autumn, there have been isolated reports of adult dragonfly sightings in December from Maryland and Delaware, and there is even an early January record for New Jersey, these are extremely rare. Normally, the last adult dragonflies disappear with the hard freeze that typically occurs well before Thanksgiving. By January, ponds may be ice-covered, and air temperatures occasionally drop below 10°F. Schoolchildren find it amazing that the larvae of dragonflies and damselflies are alive and well in the water beneath the ice, which never drops below 32°F.

However, Delmarva bays, a type of vernal pond common on the Coastal Plain of the Delmarva Peninsula, represent an interesting challenge for dragon-flies that prefer these fishless habitats. Whether or when these ponds re-form and to what depth after a dry summer makes living there risky for a larva that will be over-wintering.

One Delmarva bay inhabitant is the larva of the Blue-faced Meadowhawk, that becomes a particularly attractive smallish dragonfly with a china-blue face, a brownish thorax, and a pink-red abdomen with dark marks down the side. Blue-faced Meadowhawks seem to know which depressions will fill with water over the winter after a dry summer. They gather there in late summer and early fall to mate and lay eggs, relying on rains, snow melt, and reduced evapora-tion to refill the ponds. This usually happens, but occasionally, as in the winter of 2001-2, it did not, and many of these depressions lacked standing water for more than a year. Following that year and for several years thereafter, Blue-faced Meadowhawks were hard to find. This is yet another example of how physical conditions affect the abundance of a species [88-95].

The larvae of other species that live in ponds fed by flowing water where water levels fluctuate little have more stable populations. However, they must contend with fish, predators that Blue-faced Meadowhawk larvae do not en-counter in Delmarva bays. Each habitat has its advantages and disadvantages, and natural selection has produced adaptations that permit all species of drag-onfly and damselfly to deal with their preferred habitat.

Blue-faced Meadowhawks emerge anywhere from late June through August and are most common in September into October. We do not know how they deal with a pond that dries up early. In some other species, larvae are known to burrow in the mud and wait a year before emerging. Others may accelerate their development and emerge before the water is gone. Selective pressures in this type of environment may have favored individuals of a few species in other parts of the world to tolerate moist, terrestrial environments, which the species now use instead of their original aquatic habitats.

This mating pair (above left) and male Blue-faced Meadowhawk (above right) were photographed in mid September at the Whale Wallow, a Delmarva bay in Lums Pond State Park. At the time there had been no standing water in the depression for over two months. In December under the ice of the Whale Wallow (below right) the larvae of the Blue-faced Meadowhawk are alive and well. They will emerge in late June and July before this Delmarva bay dries up.

77 *WESTERN VAGRANT
Sympetrum corruptum (Variegated Meadowhawk)

The Variegated Meadowhawk is an example of a dragonfly that has not yet been found on the Delmarva Peninsula, but is quite likely to occur here. Records of the Variegated Meadowhawk exist for Sable Island, a sandy spit of land that is part of Nova Scotia and 100 miles out to sea in the Atlantic Ocean. Every few years one shows up somewhere along the East Coast or near the eastern Great Lakes. However, no one has ever reported one for the Delmarva Peninsula. Clearly, that is an oversight of a dragonfly that migrates far and wide in the western United States. Eventually one will catch someone's eye as it passes through.

By chance, I documented the first and only known record of the Variegated Meadowhawk for Maine. While vacationing with my family in Acadia National Park and enjoying a day at Sand Beach, I took a few moments to explore the warmer freshwater marsh behind the dunes. The presence of a Variegated Meadowhawk immediately caught my attention and created a dilemma. I had obtained a collecting permit, but it stipulated that I could not

HAL WHITE

A young male Variegated Meadowhawk showing a yellowish straw-like coloration. As it ages, it becomes much redder.

collect conspicuously where there were park visitors. Thus, I had left my net in the car.

For the next half hour, with my youngest daughter on my back in a Gerry backpack, I followed this reddish-orange and brown, medium-sized dragonfly from perch to perch trying to get close enough to catch it with my fingers. Frustratingly, every time I got close, it would fly off to another perch, and I would follow. Finally, I succeeded in catching it and can attest to how difficult it is to catch a dragonfly without a net. Although I didn't have a net to advertise my activities, surely a 6′ 4″ man in a swimming suit with a child on his back, crawling around in the sand to creep up on a dragonfly must have attracted the attention of some park visitors. Nevertheless, I didn't lose my permit. The specimen now resides with other specimens in a collection at the park's museum.

Freshwater or slightly brackish ponds along the Delmarva coast would be the most likely place to discover the first Variegated Meadowhawk. While it might even breed at such sites, the next generation would move on to another place. Although its season is long in the West, records for the East tend to be in the late summer and fall suggesting they have flown a long way if they started west of the Mississippi. It is undoubtedly the most common dragonfly in the western half of the United States. Driving down a road for mile after mile, one notices that barbed-wire fences by the road will sometimes have one perched on every barb. Sometimes Variegated Meadowhawks migrate in huge numbers along the Pacific Coast. Seeing one there is nothing special, but seeing one on the Atlantic Coast definitely is a notable event.

*The Varigated Meadowhawk is one of three species included in this book that have not yet been confirmed to occur on the Delmarva Peninsula.

78 WHEN BINOCULARS DON'T HELP

Sympetrum janeae (Jane's Meadowhawk)
Sympetrum rubicundulum (Ruby Meadowhawk)

Sid Dunkle's book, *Dragonflies through Binoculars*, stimulated many birders to develop an interest in these large and beautiful insects (*36*). In fact, so many have become interested that the old-timers who spent years in the field alone collecting specimens with a net, who learned the scientific names of the species before the "common" names were proposed, and who looked at specimens under a dissecting microscope, are becoming an endangered species. Those who have used Dunkle's book can certainly admire the beautiful pictures, but they soon realize that matching pictures with a specimen seen through binoculars (or in the hand) is not always easy and sometimes is impossible. There are several examples in this region of separate species that can only be distinguished by catching a specimen and examining it with a hand lens or microscope. The Ruby Meadowhawk and Jane's Meadowhawk represent one such species pair.

Males of both species are bright red with small triangular black spots on the sides of the abdomen. They cannot be distinguished by size, they occur together at the same wet meadows, and they have similar flight seasons starting in June and extending into autumn. While experts agree they are separate species, I know of no one who can confidently distinguish one from the other without a close look at the male hamules, the male sex organs on the underside near the base of the abdomen.

The famous American entomologist, Thomas Say [**91**], described the Ruby Meadowhawk in 1839. More than 100 years later in 1943, B. Elwood Montgomery decided that really there were two species, and described a Midwestern species, now called the Cherry-faced Meadowhawk (*Sympetrum internum*). Then in 1993, Frank Louis Carle described an eastern species, Jane's Meadowhawk (*22*). However, experts debate whether Jane's Meadowhawk is just an eastern form of the Cherry-faced Meadowhawk without a cherry-red face. What is certain is that there are two species on the Delmarva Peninsula, and they are indistinguishable through binoculars. To further complicate matters, even the dragonflies may not always make the distinction, and some hybridization may occur, although this remains to be demonstrated for certain. As much as some people despise collecting, it is impossible to document these species without a specimen in hand.

Rather than get tangled in taxonomic knots that in time may be untied, it is better to appreciate these two closely-related and colorful species for what they are and what they symbolize in the march of seasons. The Japanese capture that in a wonderful form of poetry called haiku in which red dragonflies feature prominently (*113*).

> The beginning of autumn,
> Decided
> By the red dragon-fly.
>
> SHIRAO

It is difficult, if not impossible, to distinguish the Ruby Meadowhawk from Jane's Meadowhawk without a hand lens or a microscope. Because the individual was not captured and examined closely, its identity is not known.

79 SPIDER WEBS
Sympetrum semicinctum (Band-winged Meadowhawk)

Some tropical damselflies apparently feed on spiders that they pluck from webs. To have such a specialized diet implies the ability to recognize spider webs. Once in California I observed a skimmer dragonfly on a collision course with an orb spider's web. Just inches in front of the web, it stopped, hovered, dropped down a foot or so, and flew under the web (*99*).

On the other hand, many times I have observed dragonflies, sometimes rather large ones, caught in spider webs. Occasionally webs will have more than one victim. This makes me wonder why the dragonflies were not able to avoid the web and how they got caught. Was it early or late in the day when the lighting was poor? Were they trying to escape a bird or involved in a territorial chase? Are there some species that recognize and avoid webs and some that do not? Are there some spider webs that are harder for dragonflies to see and thus catch more dragonflies? …or stronger, or designed differently, or placed differently?

These are just a few of the interesting questions a curious middle school or high school student could explore with a summer-time science project. Perhaps the best place for such a project would be a pond where numerous garden spiders have spun their webs in the surrounding tall grass. A survey of webs once or twice a day might reveal interesting patterns.

Several times spider webs have informed me of dragonfly species that I had not seen and thus would have otherwise overlooked in a survey. Among the species that I have found in a spider web is the Band-winged Meadowhawk. The amber brown on the basal half of its wings makes it distinctive and easy to identify even when well-wrapped and distorted by a spider. The Band-winged Meadowhawk differs from other meadowhawk species in that it has a distinct preference for wet meadows associated with flowing spring water on stream flood plains. So far, it is only known north of the Chesapeake and Delaware Canal. However, the Band-winged Meadowhawk is not restricted to the Piedmont region, so one might encounter it farther south.

Band-winged Meadowhawks emerge during June and fly all summer. They are uncommon in this area. The next time you see a spider web near a wet meadow, look closely and perhaps you may find a Band-winged Meadowhawk or some other dragonfly caught in it.

Spider webs near wetlands pose significant hazards to dragonflies such as the male Band-winged Meadowhawk (top) found in Maine. The bottom photograph of a male Band-winged Meadowhawk was taken in Delaware.

80 DELMARVA DRAGONFLIES IN DECEMBER

Sympetrum vicinum (Autumn Meadowhawk)

By Thanksgiving, the weather has turned cold, the trees have lost their leaves, and there might even have been some snow. Most dragonfly enthusiasts have put away their gear more than a month earlier. Yet, there may still be Autumn Meadowhawks around. These small, red-bodied dragonflies seem to defy cold weather. As the dragonfly season fades but stretches with a milder than normal autumn, enthusiasts in the Northeast seem to vie to be the last to see a live dragonfly in the wild before the first severe cold snap. They report their sightings on the Odonata ListServe. On the Delmarva Peninsula there are reports of this species as late as December 8th from Salisbury, Maryland, and even later in southern New Jersey. Glimpses on December 5th in southern New Hampshire and December 14th in northern Ohio are impressive.

Philip P. Calvert, one of the most prominent dragonfly experts of the last century, also found this late season search enjoyable, although he didn't have the Internet to proclaim his sightings to the world. He waited more than 25 years to summarize and publish his observations (*18*). As a professor of biology at the University of Pennsylvania, he would take a three-minute stroll between classes to a small pond in the campus Botanical Garden where he would hope to see the tell-tale flash of red among the brown leaves, indicating that the Autumn Meadowhawk, *Sympetrum vicinum* as he would have called it, was still around. In 24 years of observations between 1898 and 1924, his latest date was November 23rd, sometime after the first hard frost.

Interestingly, Autumn Meadowhawks do not wait until fall to emerge. They first appear in June and usually display the yellow color of young adults when found. They seem to be scarce all summer and then become quite common in early fall. Males flying in tandem with the brownish-red females are particularly noticeable over shallow water as the females tap the water to lay their eggs.

Autumn Meadowhawks know how to stay warm on cool days. They gather close to the ground on sunny south-facing slopes that are protected from the wind (*50*). We don't know what they do in the evenings to stay warm or whether they, like some other insects, produce molecules that serve as antifreeze when the air temperature drops below freezing. Some nice science projects await some curious and enterprising students here. Are the later dates now due to more observers, better habitats, or a warmer climate than Calvert experienced nearly a century ago?

The male Autumn Meadowhawk (top) has a bright red abdomen which is often seen among the fallen leaves in the fall and occasionally even after Thanksgiving. When it emerges in the summer, it is yellow (below left). The female (below right) is perched on the bark of a tree in the sun to warm up.

81 COASTAL TRAVELER
Tramea calverti (Striped Saddlebags)

Others share my fascination with weather. Habitually, I will look at weather radar when there is no rain in the forecast, the sky is cloudless, and there is not a drop of rain within 300 miles. In a way, my obsession was rewarded by a message on the Odonata ListServe from Bob Barber, then living in New Jersey. He called attention to a weather phenomenon, familiar to beach lovers, which is sometimes observable by radar on clear days, and its possible implications for dragonflies.

On hot summer days the land and the air above it warm rapidly, while the ocean water remains relatively cool. As the day progresses, the warming air, being less dense, rises in thermals that form puffy fair-weather clouds over the land. The cooler and denser air over water moves inland to replace the rising air, keeping the shoreline free of clouds and resulting in the refreshing sea breeze that beach lovers enjoy. Normally, sea breezes move inland only a few miles. That boundary is sometimes visible on radar as a long, snaking thin line

None of the other saddlebag species have the prominent thoracic stripes like the Striped Saddlebags.

JIM WHITE

parallel to the coast. Why is it visible by radar? There must be something that reflects the radar beam.

Bob Barber's suggestion is that this narrow band of converging air concentrates small flying insects and the dragonflies that feed on them. The radar detects this aerial debris. Furthermore, the narrow band creates a food corridor for migrating or dispersing dragonflies. This idea follows from the fact that tropical species living in seasonally dry areas must find breeding sites. They apparently do so by flying high into the air streams that converge to produce rain hundreds of miles away.

The Striped Saddlebags, a tropical species that invades the Northeast along the Atlantic coast from time to time, may employ this strategy and follow the coastline corridor northward. In 1992 Striped Saddlebags appeared in a number of coastal locations from Cape May, New Jersey, to Central Park in New York City. Subsequently, they have appeared as far north as Cape Cod. Although breeding and oviposition was noted in southern New Jersey (79), it is unlikely that the individuals observed had emerged locally or that their offspring would survive the winter.

To get to Cape May or Cape Cod, Striped Saddlebags undoubtedly traversed the Delmarva coastline on their way north, but most of the time escaped notice. Our first record was a female collected by Clark Shiffer in Ocean City, Maryland, on August 22, 1976 (76). In August 2010, Jim White encountered this species in Kent County, Delaware, a new record for the state. It is mostly red with a yellow face, two white stripes on each side of the thorax, and a black tip to its abdomen. The wings veins are red but only a stripe near the base of the wings next to the body between the veins is red. This red area is much less extensive than for the Carolina Saddlebags [82]. When you are at the beach, be on the lookout for a Striped Saddlebags flying by.

82 COLONIAL WETLANDS HAVE CHANGED

Tramea carolina (Carolina Saddlebags)

In one's youth, the natural world seems permanent. Only after many years, when familiar landmarks have disappeared and fields where you picked berries are now shady woods or housing developments, does the reality of impermanence assert itself. With that awareness and a little imagination one can speculate about the past habitats and fauna of the Delmarva Peninsula. It is then one starts to struggle with the conceptual meaning of "natural areas" and "restoration" with respect to wetlands in this part of the world.

Look at a road map of the Delmarva Peninsula and locate the large ponds. None are natural. With almost all of them, a stream was dammed to support a water-powered mill. To dragonflies and damselflies of colonial times, those ponds were like gigantic beaver ponds and represented types of habitats that hardly existed before. The Skimmers and their relatives undoubtedly benefited. At the same time, the dams and mill ponds flooded the swiftest stream habitats, the favored home of many Clubtails.

Look at a smaller scale topographic map and note the many small depressions that contain water in the spring and sometimes longer, depending on rainfall. These so-called Delmarva bays or Carolina bays were a nuisance to farmers; thus, many were drained. With that went the species that thrived in fishless vernal pools. Similarly, parts of southern Delaware and the Eastern Shore of Maryland have drainage ditches that mark where swamps used to be. The present-day corn and soybean fields there provide no home for dragonflies.

There are no field guides to the dragonflies and damselflies of colonial Delmarva, so we don't know what Caesar Rodney, a Delaware representative at the 1776 Continental Congress, might have seen flying around. While the fauna undoubtedly would have been different, we have no documentation. In fact, before 1763 no dragonfly species from what is now the United States had been described. In that year, the Swedish botanist Linnaeus described *Tramea carolina*, the Carolina Saddlebags, a species he never saw alive. This striking red dragonfly seems to prefer Delmarva bays, so one might imagine that the species was much more common then than now. At times it seems to fly endlessly and is hard to approach, while at other times it perches at the tip of a dead twig between short feeding flights. The expanded base of its hind wing, colored brown with red veins, provides the ability to glide and sustain flight. This species and its relatives often migrate on the winds and congregate in swarms over

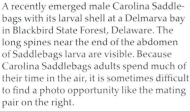
A recently emerged male Carolina Saddle-bags with its larval shell at a Delmarva bay in Blackbird State Forest, Delaware. The long spines near the end of the abdomen of Saddlebags larva are visible. Because Carolina Saddlebags adults spend much of their time in the air, it is sometimes difficult to find a photo opportunity like the mating pair on the right.

fields. The ones that show up in the early spring may come from farther south.

Just as colonial wetlands and their Odonata fauna have changed in the intervening 250 years, so will the wetlands and Odonata fauna of today change in the next 250 years for many different reasons, only some of which we can envisage. We can hope that the Carolina Saddlebags and its wetland habitats will still be here.

83 THE WING: A PLEATED NETWORK

Tramea lacerata (Black Saddlebags)

As master aerial predators of the insect world, dragonflies correspond to highly maneuverable jet fighter planes. With several hundred million years to master the air, they have evolved capabilities that continue to challenge aircraft engineers such as the transition from hovering to high-speed pursuit. Much of this maneuverability comes from the structure of their wings that have a dense network of veins and cross veins. In fact, at one time, dragonflies and damselflies were classified with the Neuroptera (net-winged insects) because of this trait.

At first glance, a dragonfly's two pairs of wings appear flat, but they are not. Closer examination reveals that the main veins extending away from the thorax form alternating ridges and valleys on the wing surface with supporting cross struts between them. This gives the wing stiffness while remaining thin and light-weight, much the same way that a floppy sheet of paper makes a stiff fan when it is folded into pleats.

The leading edge of the wing that meets the air in flight is the strongest vein and is called the costa. Each of the other major veins and branches has a name as well. The number of cross veins between various landmark branches differ among species. These taxonomic characters prove useful, though tedious, in classifying dragonflies into families, genera, and sometimes species. Most of the fossil dragonflies are known only from their fossilized wing impressions.

The Black Saddlebags, like other Saddlebags species and their Glider relatives [72-73], has very wide hind wings with a fine network of veins to support the added wing surface. This extra surface enables Black Saddlebags to glide on the wind and travel long distances. They commonly are seen flying over fields far from water. They are fairly common around ponds in the summer and are easy to recognize by the mostly black head, thorax, and abdomen, and a conspicuous black area at the base of the enlarged hind wings. They have yellow spots toward the end of the abdomen. Emergence first takes place around the third week of May and adults fly all summer. Those emerging in early October have completed a life cycle over the summer and are the offspring of adults seen earlier in the year. Most likely, they will wing their way southwards and reproduce there.

The male Black Saddlebags frequently patrols over ponds on the Delmarva Peninsula but doesn't perch often. A careful look at the wings of this male reveals they are not flat. Rather, major veins form ridges and valleys in a pleated array that stiffens the wing. This is the case for the wings of all dragonflies and damselflies.

THE DRAGONFLY SOCIETY OF THE AMERICAS

Tramea onusta (Red Saddlebags)

The number of people interested in dragonflies and damselflies has grown explosively in the past two decades. Certainly the publication of attractively-illustrated field guides that cover the country, various regions, and individual states has had a great deal to do with the new fascination. In the past, a few people with a common interest gathered once in a while for a day of collecting in some interesting place. As the numbers grew, things became more organized. Perhaps the first formal meeting of North American dragonfly enthusiasts took place in March 1963 at Purdue University [50] as part of a symposium at the annual meeting of the North-Central Branch of the Entomological Society of America. (A photo of participants follows this vignette.) B. Elwood "Monty" Montgomery, a Professor of Entomology at Purdue and the authority on Indiana dragonflies,

BOB MOUL

The Red Saddlebags is migratory from the west and occasionally will complete a life cycle in the east as this young male did in southern Pennsylvania.

organized the meeting. About 35 people met then and again three years later.

Since 1966, meetings have occurred with increasing frequency and now ten or more regional and national meetings happen each year. The Dragonfly Society of the Americas (DSA), formed in 1989, now sponsors one national meeting and about five regional meetings each year. Unlike the junket meetings for some professions in resort cities or on luxury cruise ships in the Caribbean, these meetings seek inexpensive rural motels to minimize expense and travel distance to dragonfly habitats of interest. The schedule is informal, with the group gathering every morning in the parking lot to assess the weather, look at maps, decide where to go, and form car pools. At the end of the day, the parties return to share their experiences over dinner, clean up, and enjoy slide shows and conversation into the evening.

A memorable DSA meeting for me was held in the Hill Country of west Texas; it included a post meeting trip to the Nature Conservancy's Devil's Falls Preserve. This spectacular oasis had a variety of dragonflies previously unknown to me. Among the species common there was the Red Saddlebags, a species that has been found on the Delmarva Peninsula only once or twice. It looks quite similar to the Carolina Saddlebags [82], but lacks violet on its forehead, and the sides of abdominal segments 8 and 9 are red instead of black. There are also slight differences in the shape of the red patch at the base of the hind wing.

Like other saddlebag species, the Red Saddlebags wanders far from its emergence site. Although it normally occurs west of the Mississippi, strays make it to the eastern United States from time to time. I thought I saw several Red Saddlebags in a field at Lums Pond State Park in 2008. Unfortunately, they eluded my net and camera for positive documentation.

Participants at the first meeting of North American dragonfly specialists at Purdue University in March 1963. Many of the people in this photo are mentioned in this book: George (1) and Juanda Bick (10), George (2) and Alice F. Beatty (9), Donald Borror (3), Thomas W. Donnelly (4), Minter J. Westfall (5), B. Elwood Montgomery (6), Philip S. Corbet (7), Leonora K. Gloyd (8), and Hal White (11). (Photograph reproduced with the permission of the North-Central Branch of the Entomological Society of America.)

PART II DAMSELFLIES (ZYGOPTERA)

85 SENSE OF WONDER
Calopteryx dimidiata (Sparkling Jewelwing)

Occasionally I get entangled in the seemingly futile debate about whether a scientific understanding of natural phenomena destroys our sense of wonder. Usually the "discussion" begins when someone tries to explain something such as why the sky is blue and not some other color to someone who is exhilarated by a crisp cloudless spring day when the grass is bright green, flowers are blooming, and birds are singing. The victim's sudden exit from reverie can polarize the conversation with a blunt, "I *don't care* and I *don't* want to know! You just *ruined* my enjoyment of a beautiful day!"

The line between a sense of wonder and a sense of curiosity is easy to cross. When one reads Thomas Eisner's book, *For Love of Insects* (*37*), the two gloriously merge. To paraphrase Nobelist Albert Szent-Györgyi, Eisner sees what others have seen and thinks what no one has ever thought. The wonder of what we see every day and usually dismiss is revealed in the studies of Eisner and his coworkers. For me, such understanding enhances my sense of wonder.

The beauty of the Sparkling Jewelwing can't help but elicit a sense of wonder. This fairly large, yet delicate, damselfly has an emerald green body and clear wings tipped with black, though some females lack the black tip. Sparkling Jewelwings prefer clean, sandy-bottomed streams on the Coastal Plain, where they perch on overhanging or floating vegetation and occasionally dance out

RICHARD ORR

JIM WHITE

Few dragonflies or damselflies can match the beauty and elegance of the Sparkling Jewelwing. Here a male (left) and female (right) are shown at their stream-side perches in Caroline County, Maryland.

over the water. Their flight season extends from early May to the end of August. Often their more common and widespread cousin, the Ebony Jewelwing [86] is on the same streams.

Interestingly, the Chesapeake Bay may be a geographic barrier that separates northern populations of the Sparkling Jewelwing from those farther south. On the Delmarva Peninsula the superior appendages of males in side view are curved distinctly downward, while in the south the appendages are straight (47). There is also a trend to more extensive black on the wing tips south of the Delmarva Peninsula. As with all damselflies, in contrast to dragonflies, the front and hind wings have the same shape [1].

While the colors of most dragonflies and damselflies come from pigments, the green color of the Sparkling Jewelwing comes from another source—iridescence. Sunlight striking tiny parallel ridges arrayed at regular intervals on the surface of their bodies causes interference of the light in the same way that gives peacock feathers their iridescent colors. I hope this little bit of incomplete explanation about color produced by iridescence will not interfere with the sense of wonder you experience on seeing your first Sparkling Jewelwing.

86 COURTSHIP AND CUCKOLDRY
Calopteryx maculata (Ebony Jewelwing)

Sometimes familiarity breeds contempt, but that could never be said of the Ebony Jewelwing. These are spectacular insects and a delight to watch. Of all the damselflies found in eastern North America, this one is probably familiar to more people than any other. As damselflies go, it is large and conspicuous with densely-veined, dark wings and an iridescent green (sometimes blue) body. In the males, the wings are black, while the female's wings are smoky with a conspicuous white spot near each wingtip. The Ebony Jewelwing is widespread and common. Frequently, it will be the only species seen along woodland brooks, though it has a wide tolerance for streams of different types. It is hard to miss during its flight season from early May to early September. The genus name, which means beautiful wings, says it all.

Imagine sitting on a rock in the middle of a crystal clear, spring-fed woodland stream on a cloudless day in June. The air is fresh. Shafts of sunlight dance on the water and on the streamside vegetation. Perched around you on the sunlit leaves is a congregation of over twenty male Ebony Jewelwings. They occasionally flick their wings or fly out over the water. You are an intruder in their territory, but if you remain still you may notice females come flapping up the stream with a characteristic bounce. While you may be ignored by the males, females cannot pass by unnoticed by the males.

This idyllic setting is where males encounter and compete for females. The competition is winner takes all. If a female has mated recently, the next male to copulate will remove the previous suitor's spermatheca (a packet of sperm) from the female and replace it with his own (*86*). No wonder the male often escorts the female to her oviposition site on floating leaves to assure fatherhood. Such discoveries by behavioral ecologists have become possible with the development of microscopic and molecular techniques used in the laboratory coupled to good field observations.

I grew up with a cold-water stream in my backyard. Ebony Jewelwings were often the only species of Odonata that one could expect to see every summer day. With black wings, the males were quite conspicuous as they flew along the edge of the stream. Before they acquired the name Ebony Jewelwing, a friend of mine liked to call them "Flapdoodles," because of the bouncy, characteristic flapping flight they had. We had no idea of the high-stakes biology we were witnessing.

Ebony Jewelwings are common on Delmarva streams having noticeable flow. The green, metallic color on the abdomen is due to iridescence. The females (right) can be distinguished from males (left) by the white spot near the ends of their smoky black wings.

87 FALL COLORS
Hetaerina americana (American Rubyspot)

Growing up, I thought there was only one kind of firefly—the ones my friends and I caught by hand and put in jars on warm summer nights. Long before I had heard about the Q_{10} Rule (on average, the rate of a biological process doubles for each 10°C increase in temperature) or Arrhenius Plots, my parents said I could determine the temperature by measuring the time between their flashes—something that seemed pretty neat—but I never did. Later I learned about different species of fireflies and how hungry females of one species could lure unsuspecting males of other species to their death by mimicking the flash patterns the males recognized as their own species. Even later I learned from firefly taxonomists that firefly species are almost impossible to distinguish without knowing the flash pattern of the living organism.

It seems that problems related to the loss of field characteristics in dead specimens confront all taxonomists whether they work on fireflies or some other group of organisms. Taxonomists try to find anatomical characters that reliably distinguish museum specimens of one species from those of another. The lengths, widths, ratios, or shapes of anything measurable are fair game. Colors,

The female American Rubyspot (right) has the iridescent green body but not the bright red spot at the base of the wings that characterizes the male (left). Look for American Rubyspots in late summer along Piedmont streams. They perch in the sunlight at the tips of streamside vegetation that bend toward the water. Despite their bright colors, they are hard to follow in flight.

though used by dragonfly taxonomists, are often unreliable because they frequently change during maturation, are hard to describe accurately, and usually fade quickly after death.

Sometimes I am amazed at how difficult it is for me to identify color-faded specimens that someone else has caught when the specimens contain no information about habitat, location, or the time of year. For an experienced observer of live dragonflies in the field, such information as color, habitat, behavior, and time of year enable instant identification. Under the microscope, a taxonomist must rely on different characters and information to identify specimens.

In the northeastern United States, no one could mistake an American Rubyspot for anything else, in the field or under the microscope. Males have a red thorax and iridescent bronzy body. Each densely-veined wing has a brilliant red area at its base. Females also have iridescent bodies, but green replaces red, and the wings are clear or slightly colored. In contrast to regions not far to the southwest, where this species flies from spring to late fall, in our area we don't expect to see the American Rubyspot before late July. They fly until a hard frost. Look for them perched or flying close to the water along the banks of medium-sized Piedmont streams with noticeable flow. There are a few records from the Coastal Plain of the Delmarva, where they add their color to the colors of autumn.

88 MOVING EAST?
Archilestes grandis (Great Spreadwing)

I have yet to figure out the Great Spreadwing, the largest damselfly found in the northeastern United States. I first encountered an apparently thriving population in mid-October along a bog stream at over 1800 feet elevation in central Pennsylvania, a decidedly unexpected habitat. That ephemeral population disappeared within a year and has not been seen there since.

Much to my delight ten years later, the very first damselfly I encountered upon moving to Delaware was the Great Spreadwing. I found a male perched on a tiger lily stem by my front porch in late August. Over the next couple of weeks I observed about ten more around the house and nearby along a small drainage ditch through a lawn on the University of Delaware campus. That turned out to be nature teasing. In the subsequent three decades I have seen fewer than 5 individuals, usually singly near small, slow-moving streams.

E. B. Williamson, an Indiana banker who was also one of the most noted American dragonfly specialists of the early 1900's, discovered a population of Great Spreadwings along a small polluted creek in a mowed area on a college campus in Oxford, Ohio in 1927. At that time when the only records were from the West Coast, other than a single Arkansas record, he proclaimed, "It is altogether the most surprising find I have ever made in collecting dragonflies" (*106*).

Because the Great Spreadwing has a very late flying season and lives in habitats not often explored for dragonflies and damselflies, it may be more common than it appears. Due to the comments of Williamson and others, it is known as one of a few species that have greatly expanded its range from the southwest [106]. While the absence of records for the eastern United States before 1920 may be evidence for such, my experience suggests that it may have been overlooked. As they say, "absence of evidence is not evidence for absence".

Regardless of whether or not the Great Spreadwing has moved here in the recent past or just has a life cycle and preferred habitat that keeps it out of contact with people carrying nets, it is a magnificent damselfly to find. Compared to most damselflies, it has a robust body that is two or more inches long. The two bright-yellow stripes on the sides of its thorax make for easy identification. Though its flight season starts in July, it seems to be easier to find later in the summer.

At a little over two inches in length, the Great Spreadwing is the largest damselfly found on the Delmarva Peninsula. Photograph taken in central Pennsylvania.

TRIBUTE TO EDMUND M. WALKER
(1877-1969)

Lestes australis (Southern Spreadwing)

When I first got interested in dragonflies and damselflies, there were no field
guides. Fortunately, I had experts who introduced me to the scientific literature. Through that experience, I came to appreciate the difference between
primary and secondary sources, and I became familiar with the names of past
and present leaders in the study of these insects. The names included James G.
Needham [**65**], Minter J. Westfall, Jr. [**111**], Philip P. Calvert [**108**], Hermann
Hagen [**107**], E. B. Williamson [**52**], Leonora K. Gloyd [**101**], Donald Borror
[**50**], George and Juanda Bick [**97**], B. Elwood Montgomery [**84**], Clarence H.
Kennedy, and Edmund M. Walker. Of these, the works of Walker, a prominent
Canadian entomologist (*101*), captured my attention the most. Walker had written the books on *The Odonata of Canada and Alaska* (*88, 89, 91*), and he was the
world authority on *Aeshna* (Mosaic Darners), a group of large colorful dragonflies [**2-4**] that I had encountered in large numbers on canoe trips in Canada. He
described their behavior and habitats in much greater detail than other writers

This pair of Southern Spreadwings looks very much like their close relative the Sweetflag
Spreadwing [**92**]. Aside from a close examination of the male terminal appendages, the two
can be separated by comparing the relative length of the second abdominal segment to the
third in males. If it is less than half the length, it a Southern Spreadwing. This trait is mirrored
in the female ovipositor which must engage the second segment of the male during mating. For
the Southern Spreadwing, the ovipositor does not extend beyond the end of the abdomen.

I had read. Perhaps having fewer species to deal with in Canada, Walker could pay more attention to each one.

It was Walker who recognized a vexing taxonomic problem and tried to sort out the relationship among three very similar-appearing spreadwing damselflies. He concluded that the Southern Spreadwing and the Common Spreadwing (*Lestes disjunctus*) were subspecies while the Sweetflag Spreadwing [**92**] was a separate species despite its superficial similarity with the Southern Spreadwing (*87*). DNA analysis seems to indicate a subspecies relationship. However, others have recently concluded that all three are distinct species (*30*). It was Walker who recognized and described the Southern Spreadwing. In our area only the Southern and Sweetflag Spreadwings occur, but they are hard to tell apart without a hand lens and a good idea of what to look for. The Southern Spreadwing is the first spreadwing to emerge starting in early May. Later on in the season, when both can occur together, the separation is difficult.

The Southern Spreadwing usually occurs at fishless ponds such as Delmarva bays. A drought in 2001-2002 dried up most of these sites for an extended period and had a devastating effect on populations. Only recently have they reappeared at a few places.

90 DEALING WITH DROUGHT
Lestes congener (Spotted Spreadwing)

Although the eastern United States experiences extended periods of dry weather, there is no annual dry season; a month without measurable precipitation is quite rare. Nevertheless, many small streams and vernal ponds normally dry up each summer. The life cycle of the Spotted Spreadwing seems to be adapted to this seasonal cycle (*71*).

The Spotted Spreadwing ranks as our smallest, least colorful, and most cold-tolerant spreadwing. It is common in the northeastern United States, but its known distribution is restricted to the more northern counties of the Delmarva Peninsula. It typically emerges in late June or early July when temporary ponds are about to dry up. It goes relatively unnoticed for most of the summer; only after the first frosts have arrived and most leaves have fallen does it become a prominent member of the surviving damselfly community. My latest record is November 10th, but it likely flies until the first hard freeze. There are records in December for New Jersey. Aside from being smaller than other spreadwings and having brown markings, both males and females are easily recognized by the pair of brown spots on the lower sides of their thorax.

A good place to look for the Spotted Spreadwing is among the wetland vegetation in depressions that hold water for half the year or more. In the late summer and fall when the water has dried up completely, the spreadwings seem to know the water will return. Spotted Spreadwings come back to lay their eggs in living, or more often, dead stems. As shown in the picture, the male often escort the female as she inserts her ovipositor into plant stems. Where spreadwings are common, the stems show considerable damage from the numerous punctures. After the plants die and become submerged, the eggs remain dormant until spring and the larvae later thrive when the water level is normally high.

While spreadwings seem to be adapted to ponds that dry up for part of the year, we do not know whether the eggs can survive in a dead stem for a year. The winter of 2001-2 and the following year were very dry. Several Delmarva bays that normally fill with water remained dry and stayed dry until the following winter [76]. For several years since that drought, Spotted Spreadwings have been conspicuously rare or absent at these habitats where they formerly appeared regularly. This suggests that their eggs cannot survive a dry year, or possibly that the eggs hatched during a brief damp period and the larvae did not survive.

The female of this pair of Spotted Spreadwings is ovipositing in a living stem. More often they lay eggs in dead vegetation in and around temporary ponds. In contrast to male dragonflies that clasp females by the head, male damselflies clasp females by the front of the thorax. By attending the female during oviposition, the male assures that no other male will mate with her before she lays eggs.

91 TRIBUTE TO THOMAS SAY (1787-1834)

Lestes eurinus (Amber-winged Spreadwing)

In the 1820's, Thomas Say published *American Entomology*, the first book on the insects of North America by a North American author (*83*). A native of Philadelphia and a founder of the Academy of Natural Sciences of Philadelphia in 1812, Say frequently is called the father of American descriptive entomology. Among many other species, Say was the first to describe many species including the American dog tick, the Colorado potato beetle, and the mosquito that transmits malaria in North America. In all, he described over 1500 insects. He also described many mollusks and a few mammals, including the coyote.

HAL WHITE

The Amber-winged Spreadwing was first described by Thomas Say, an American naturalist of the early 19th century. This large damselfly is now known from only one site in New Castle County, Delaware.

Thomas Say was a private person who found contemporary society unsatisfactory. In 1825, he moved to New Harmony, Indiana, to become part of a utopian community there. The community failed, but he continued to live there until his death working in less than ideal conditions for a naturalist who once had access to the facilities of the University of Pennsylvania and the Academy of Natural Sciences.

One of the species described by Say and published posthumously is the Amber-winged Spreadwing. This is one of our largest damselflies. As implied by its name, amber-colored wings distinguish it from all other damselflies in this area. Its thorax has a distinctive blue-green color, and there is a splash of yellow on the sides of the thorax. It is most often found at fishless ponds and can tolerate acidic sphagnum pools in bogs. Because of my familiarity with the species at a large pond in central Pennsylvania, I associate it with spatterdock in which females oviposit.

In the mid-1970's, single individuals of the Amber-winged Spreadwing showed up twice on the Delmarva Peninsula; both times were in New Castle County, Delaware, in temporary small pond habitats only a couple of miles apart. One pond was in the ore pit at the top of Iron Hill. The other was a disturbed Delmarva bay in a field under a power line near the Maryland border. Unexpectedly, in 2009 I discovered a small population at a deep clay-bottomed, fishless pond in Delaware's Blackbird State Forest.

92 COSTS AND BENEFITS OF AVOIDING FISH

Lestes forcipatus (Sweetflag Spreadwing)

Life as a dragonfly or damselfly larva is tough. Most larvae do not make it from their immature life in the aquatic world to adults in the aerial world. Despite being formidable predators in their own right, they are not at the top of the food chain. As any serious fisherman (the top predator) knows, dragonfly larvae are good as live bait. In the underwater world, fish feast on dragonfly and damselfly larvae. With such dangers, one can see why some species have adapted to life in fishless ponds.

To survive, larvae depend on camouflage. Unlike the colorful adults they may become, they are drab and cryptically colored to camouflage themselves. Lurking motionless beneath mud surfaces, in dead plant litter, and among forests of aquatic vegetation, they wait to ambush smaller insects and occasionally very small salamanders or fish. They move slowly and deliberately. Many play "possum" when handled. Certainly, larvae lucky enough to have hatched from eggs laid in a fishless pond escape the threat of fish predation. However, that benefit is not risk-free.

Typically, fishless ponds are temporary or seasonal wetlands that depend on precipitation and the fluctuating water table for water. Being isolated from streams that could maintain the water levels, many ponds dry up each summer, although some larger and deeper ponds may hold water for several years at a time. The so-called Delmarva or Carolina bays, which dot the mid-Atlantic Coastal Plain, provide habitat for many dragonflies and damselflies that occur infrequently elsewhere. Sometimes relatively new farm ponds are fishless and will host these species, but that utopia ends when the bass, perch, and bluegills arrive.

This situation illustrates the benefits, costs, and trade-offs that shape the evolution of a species' ecology. The evolution of these specialists, or non-specialists, of temporary ponds would be favored by greater survival of larvae in the absence of fish. It would be constrained by the ability of ovipositing females to recognize temporary ponds as such, and by larval ability to survive drying of the pond. Seeing such habitat specialists, biologists would look for anatomical, physiological, and behavioral characteristics that help individuals beat those challenges.

Among such adaptations in Odonates using temporary ponds might be the ability to burrow and be dormant in the ground until water returns, to develop rapidly and complete a life cycle while water is present, or to lay eggs that hatch only when water returns. Sometimes those tactics do not suffice. As noted before, the drought of 2001-2002, when many Delmarva bays remained dry through the winter [**76, 90**], illustrates the risk of laying eggs in such places. Local populations of several species, including the Sweetflag Spreadwing, disappeared. Only recently have they recolonized some of these habitats, perhaps from individuals that survived in permanent ponds in the presence of fish.

There have been times when I have seen huge numbers of Sweetflag Spreadwings fly up as I walked through emergent grasses around a temporary pond. At such times, scars from their oviposition in the stems of these grasses were readily visible upon examination.

Mature males have a blue-gray pruinosity. Females are distinguished from other spreadwings by an ovipositor that extends beyond the end of the abdomen. Adults normally emerge in June and fly into early fall.

Sweetflag Spreadwings were common in June 2000 at the Tybout Delmarva Bay in Blackbird State Forest, New Castle County, Delaware. At the time, the water was about three feet deep as it was the following year. But two years later on 1 June during a prolonged drought, the bay was dry and had been dry most of the winter. The population of Sweetflag Spreadwings disappeared and had not repopulated the bay as of 2010. However, it has repopulated a Delmarva bay in Lums Pond State Park.

93 JAWS THAT SNATCH
Lestes inaequalis (Elegant Spreadwing)

Most people consider adult dragonflies and damselflies colorful and attractive insects—even suitable models for jewelry. By contrast, most people seeing the dull-colored larvae (a.k.a. nymphs) would consider them ugly and repulsive, in a league with spiders. However, despite their looks, even the larger dragonfly larvae are harmless to humans. Closer inspection by the curious reveals that these fearsome underwater predators have impressive adaptations that put other insects, minnows, and even small salamanders at risk. Their mode of attack is ambush. Their dull colors serve to camouflage them as they lie in waiting.

Imagine your arms in front of you fused together from your elbows to your wrists with your hands side-by-side, palm up, covering your face except your eyes. You sit perfectly still eyeing a tasty morsel as it comes your way. When the prey is within range, your arms shoot out to full extension, grab it, and retract, putting the food directly to your mouth. The double-hinged lower jaw (labium) of all dragonfly and damselfly larvae works pretty much that way. Some species can snatch prey more than half an inch away. During metamorphosis to an adult, the larva loses the double-hinged jaw. The adults may also snatch prey with their mouth parts, but more often ensnare prey from the air with their spiny legs.

Among damselfly larvae on the Delmarva, members of the spreadwing family have rather long and slender lower jaws. Their labial palps, which correspond to human hands in the above analogy, have species-specific hairs (setae) and shapes that help identify them under a microscope.

The Elegant Spreadwing is one of the largest spreadwings in our area. It prefers shallow ponds with plenty of emergent vegetation. The adult abdomen and much of the thorax have a characteristic metallic green color which is sharply delineated from the yellow on the thorax and with yellow markingss elsewhere. The adult male's abdominal appendages, which are used to grasp the front of the female's thorax, are like no other in that the lower (inferior) ones extend beyond the upper (superior) appendages. This unequal feature gives rise to part of its scientific name.

Elegant Spreadwings begin to emerge in late May and adults fly into August. They may be the only spreadwing found at a pond, but usually in low numbers.

A male Elegant Spreadwing (top) and female (bottom) display their emerald green iridescent bodies as they perch in marshy grass.

94 TENERALS
Lestes rectangularis (Slender Spreadwing)

Many school children have seen the transformation of a milkweed caterpillar into a gold-studded emerald chrysalis and then into a spectacular orange and black Monarch butterfly. One of the experiences that sparked my interest in insects was seeing the process of metamorphosis when I was only five. I found a huge green caterpillar with red, orange, and blue club-like projections and brought it home. Fortunately for me, my mother knew what it was and put it in a jar, where I watched it spin a cocoon. The excitement of watching the emergence of an enormous and gorgeous Cecropia moth the following spring remains one of my fondest childhood memories. Later as a teenager, I raised so many Cecropias that the caterpillars practically defoliated my mother's lilac hedge. Such are the rewards and tribulations of parenthood. Thank you, Mom!

Anyone who has watched a butterfly metamorphose realizes that the time immediately after emergence is critical because the new adult is very soft. Even a slight touch to a wing can permanently cripple it at this time. Dragonflies and damselflies are no different. In contrast to butterflies and moths, they do not emerge from a pupa. Rather the larva climbs out of the water onto a secure perch and transforms into an air-breathing adult in its final molt.

The shed larval skin (exuvia) of a Slender Spreadwing was left behind on a blade of grass after the last molt to an adult. There is no pupal stage. The male Slender Spreadwing is distinguished from other spreadwing species on the Delmarva Peninsula by its elongated abdomen, a characteristic not possessed by females. This is apparent by the large asymmetric loop in the heart shape formed by the mating pair above right. Adults often stay hidden in shaded vegetation near marshy wetlands.

When they first emerge, dragonflies and damselflies are called tenerals and are extremely vulnerable to all sorts of danger. Some fail to escape their nymphal shell, called an exuvia, and remain trapped until they die or are eaten. Wind may blow them off their perch before they can fly. Birds love to feast on this equivalent of soft-shelled crabs. The wake of a motorboat [18] or a sudden downpour can spell doom. It takes an hour or more for the wings to fully expand and harden enough for the first flight into the relatively safe cover of nearby trees and shrubs.

Slender Spreadwings are fairly common damselflies at marshy ponds, Delmarva bays, and slow-flowing streams. Often large numbers of them emerge over a period of several days. When that happens, wading casually through the vegetation may flush many weak-flying tenerals that flutter up. If they are lucky, they will escape hungry Eastern Kingbirds and Phoebes to spend the next week hardening, feeding, and acquiring mature colors before returning to the water.

The male Slender Spreadwing differs from all other species in this area by the remarkable length of the abdomen. It is nearly twice as long as the wings. Slender Spreadwings tend to stay hidden in shady vegetation near ponds and slow streams. They lack bright colors, but close examination reveals a very attractive damselfly with subdued blue eyes, and pastel blue-green and yellow thoracic markings. The wing tips of the male are faintly highlighted in white. Tenerals first appear in late May and June. Mature adults may persist into early October.

95 SWIMMING LIKE A FISH
Lestes vigilax (Swamp Spreadwing)

Dragonfly larvae escape predators with jet propulsion [63]. They take up water into their abdomen and then expel it rapidly as they shoot forward. Damselfly larvae cannot do this. However, they have another form of rapid escape. Unlike dragonfly larvae, which have rigid, stout bodies, damselfly larvae have long slender bodies with three flat paddles known as caudal gills attached to the end of the abdomen. In addition to being highly vascularized to exchange oxygen, the gills also function like a fish's tail.

Among damselfly larvae, those of the spreadwing damselflies have truly mastered the art of swimming like a fish. A spreadwing larva can move surprisingly fast through water in short spurts by rapidly moving its abdomen from side to side and becoming streamlined by holding its legs back against its body. The larvae of other damselflies can swim this way also, but they are no match for the generally larger and speedier spreadwing larvae. I well remember the shrieks of excitement when a group of fourth graders first saw a spreadwing larva take off through the water in a porcelain pan around which they had crowded. It is quite impressive, even to adults.

In our area, spreadwing damselflies are most common at fishless ponds [92]. But unlike most other spreadwing species, the Swamp Spreadwing does not seem affected by fish. I associate it with the heavily vegetated shore of permanent ponds. Perhaps its larvae can swim fast enough through marshy vegetation to get away from fish or conversely, remain quite still to avoid detection.

The adult Swamp Spreadwing is large—up to two or more inches long. It is quite slender with emerald green metallic colors on its thorax and abdomen. The stripes on the front of the thorax are reddish brown. The tip of the abdomen becomes light gray in mature males.

The Swamp Spreadwing is relatively uncommon on the Delmarva. Although there are a few records from the northern counties, most come from ponds in southern Delaware. The flight season extends from early June to early September.

The Swamp Spreadwing looks a bit like the Elegant Spreadwing and sometimes both are found in the same habitat. The Swamp Spreadwing is slightly smaller and does not have yellow on its thorax.

96 HOME SEEP HOME
Amphiagrion saucium (Eastern Red Damsel)

The most familiar dragonflies and damselflies tolerate a variety of habitats, but they prefer common habitats such as artificial ponds. Conversely, other species are much less familiar because they have narrow habitat preferences that are rare or visited infrequently by humans. The Eastern Red Damsel falls into the latter category. Aside from being small and fairly inconspicuous, its home is away from most ponds and streams. It prefers grassy spring seepage areas and wet meadows.

A seep provides an interesting home. Because the water there seeps out of the ground at a nearly constant temperature year round, the damselfly larvae do not have to contend with the wide range of water temperatures found in ponds and to a lesser extent in streams. We do not know whether the stable temperature environment or something else like the absence of fish predators favors the use of seeps by Eastern Red Damsels, but that is the best place to find them.

Although male Eastern Red Damsels have a bright red abdomen and are often quite common locally, they don't attract attention because they spend much of their time perched low in grassy vegetation. When disturbed, they fly off to another nearby perch. They have a dark purple thorax and head with black markings on the segments near the end of the abdomen. Females are brownish and more subdued in color.

My favorite place to look for Eastern Red Damsels is in a wet meadow among the numerous seeps and trickles that feed the tributary that enters Lums Pond in New Castle County, Delaware, from the northwest. In some years they are abundant. First emergences may be as early as the end of April, and a few adults can be seen after the Fourth of July. Some of the seeps at Lums Pond State Park are within a hundred feet of a rather extensive and carefully planned wetland mitigation project.

While the idea of creating new wetlands as a mandated trade for wetlands destroyed by various construction projects sounds like a good idea and probably is in a limited sense, to think a destroyed natural wetland habitats can be recreated or replicated borders on folly (*96*). When the wetland project at Lums Pond State Park was completed in early 1997, I marveled at the design that included deep and shallow water habitats, inlets and outlets at carefully measured heights so that water would flow in a predetermined channel, and an array of stocked fish and aquatic plants. A lot of time and thought went into its design. However, humans cannot fully anticipate Nature's response.

Not planned was the arrival of beaver that immediately dammed the outlets, raising the water level and drowning many of the carefully-designed, shallow-water habitats. As surely could be expected and as I have documented, those wetlands have become a great habitat for common dragonflies and damselflies of the type found throughout Delmarva at ponds. However, despite its natural home being so close by, I have yet to see a stray Eastern Red Damsel at this wetland.

I sometimes wonder what destroyed wetland this project was meant to re-place. Quite likely it was not a pond but a wetland with specialized habitats that provided home for less common species like the Eastern Red Damsel. It is easy to create a pond, but how could anyone create a seep?

JIM WHITE

Despite the bright red color of the male Eastern Red Damsel, it often goes undetected in the deep grass around the spring seeps it prefers, due to its small size. A willingness to get wet feet is almost a necessity to see this tiny but colorful damselfly. The female is brown and is less conspicuous.

97 A DANCER'S LIFE

Argia apicalis (Blue-fronted Dancer)

Imagine observing a single species of damselfly at a small, muddy, pasture pond in Oklahoma for at least three hours in the middle of each day on 47 consecutive summer days! That's what George and Juanda Bick did from mid-June to the end of July in 1962 (*91*). They captured 315 individuals, marked them distinctively so each could be identified later, released them, and recorded their presence and activities. A total of 191 individuals were observed at least once again at the pond. The Bicks' work resulted in a detailed description of the social life of the Blue-fronted Dancer.

The study produced interesting and surprising findings. For example, male Blue-fronted Dancers monitored at the pond for three hours stayed within a 12-foot stretch of shoreline about 75% of the time. When females were observed, they almost always were paired with males. While some males did return to the pond on successive days, most did not appear every day, and some returned after many days away. Males mated on average once in their life, surviving females mated multiple times, but not on the same day, and females usually oviposited with the male attached in tandem shortly after mating. The average estimated life span was a little over eight days, although the maximum between marking and last recapture was 33 days. Each day, the damselflies started arriving at the pond about 9 AM, and most were gone by 4 PM. Robber flies, spiders,

The male Blue-fronted Dancer, here in tandem with a female, frequently lands on woodland paths near the shores of large ponds and slow-moving streams.

and Common Pondhawks [54] accounted for more than half of the observed mortality, and none was attributed to birds. Six times females were found with an attached headless male that presumably had been eaten by a Pondhawk.

This information is much more than we know about many species. What we do know of other species is often due to other studies by the Bicks. Such information is useful. Comparing the behavior of one species with another often leads to insights into reproductive strategies and adaptation to environmental circumstances. It is sobering to think that the behavior of Blue-fronted Dancers in Oklahoma may be different in some respects than on the Delmarva Peninsula, but we just don't know it because no one has taken the time to look.

Blue-fronted Dancers are common at larger ponds and slow-moving rivers on the Delmarva Peninsula. While the more commonly observed males have a pastel blue thorax and tip of the abdomen, females usually have a mottled brown appearance, but may also have a blue thorax. They commonly perch on the ground along woodland paths at places like Lums Pond, New Castle County, Delaware, where they occur from early June through mid-October. Interestingly, for reasons that are not clear, mature males can switch reversibly from their bright blue to a less conspicuous dark color—the results of another discovery by the Bicks (10).

98 SEA-LEVEL FENS: A RARE HABITAT WITH RARE SPECIES

Argia bipunctulata (Seepage Dancer)

The Seepage Dancer looks like a member of the Bluet genus. Consequently, it can be confused easily on sight with other species. However, recognition of its restricted habitat and a close look at the male appendages will avoid confusion. The Seepage Dancer is among our rarest and most threatened damselflies because it prefers boggy, seepage habitats such as sea-level fens that are typically smaller than an acre and rare—only ten or so are known on the Delmarva Peninsula. Such habitats deter most visitors.

When development threatens wetland habitats, the general public support for preservation is limited. The following fictionalized story accentuates this point and hopefully provokes some reflection:

Growing up, Jennifer Boggs spent summers at her family's cottage in Angola, a little crossroad town in Sussex County, Delaware. She enjoyed boating on Rehoboth Bay and taking short trips to the beach. Life in "slower" Delaware was great. Thus, Jen felt especially happy when, with a degree in Wildlife Conservation, she landed a job with the Delaware Endangered Species Program not too far from her favorite haunts.

During her first year at DESP, Jen spent lots of time in the field doing species inventories on various sites around the state and writing reports on what she had found. While she could identify with ease most of the birds, reptiles, and amphibians on sight, she felt overwhelmed by the number of insects that she needed to identify. Gradually she acquired an eye for things unusual and interesting. One day her supervisor assigned her to do an inventory of the Odonata of a wetland area near Cherry Walk Creek that was reported to be "the finest example of a sea-level fen anywhere in the world" (54) and only one of four such habitats in Delaware. Fenex Development Corp was seeking approval for a large condominium and recreational complex that would include a golf course and a marina. As planned, dredging for the marina would destroy the fen, and the golf course likely would modify the fen's groundwater sources.

Jen visited the site with coworkers on a steamy, hot summer day. She imagined being exsanguinated by the thick swarms of mosquitoes and greenhead flies. Poison ivy was hard to avoid. And the mud that stuck to everything was more than knee deep in places. Chuckling to herself, she thought, "No one but a wildlife biologist could enjoy this." To her the place was incredible—sundews, pitcher plants, orchids, and interesting birds and frogs. She managed to catch a tiny dragonfly, hardly an inch long, and a bright blue damselfly. Both looked unfamiliar. *Nannothemis bella* (Elfin Skimmer, [69]) and *Argia bipunctulata* (Seepage Dancer) proved to be almost unknown in the state.

That weekend Jen visited her parents at the family cottage. She excitedly described her discoveries at the fen only three miles away and how biologically important it was. Her father, a golfer and boater, and not one to mince words said, "Who cares? … I bet there aren't five people in the whole state who have ever heard of your 'Sleazy Dancer', or whatever it is. There are hundreds of people I know who would love another golf course or a closer place to launch a boat and would never miss an inconspicuous damselfly." Jen was devastated. How could it be possible to convince anyone that the site was worth saving if she couldn't even convince her father?

Without close examination, the Seepage Dancer looks like a bluet.

99 THE COLOR PURPLE

Argia fumipennis violacea (Violet Dancer)

When biologists first describe an organism, they give it a Latin name. Normally the species name is descriptive. If one is familiar with the classical languages of Greek and Latin, the name will have meaning and thus be easier to remember by association. The Violet Dancer is one of the few damselflies for which I have included not only the species name, *fumipennis*, but also a subspecies name, *violacea*. Both names have significance, but only one has relevance on the Delmarva Peninsula.

Once there were two species. *Argia fumipennis* was a purple damselfly from the southeastern United States that had strikingly distinctive dark wings—*fumipennis* means smoky wings. In the northeastern United States there was another purple-bodied species with clear wings named, *Argia violacea*. When the geographic distributions of the two were determined, a transition zone was discovered in North Carolina where one form graded into the other. Other than the striking difference in the wing colors, the two were morphologically the same. Consequently, the two were merged into a single species in which each form was recognized as a subspecies (43). Because the smoky-winged form was de-

The male Violet Dancer has clear wings. However, related subspecies from North Carolina south have yellowish brown to very dark brown wings in Florida. Oviposition occurs in vegetation at the water surface with male still in tandem with the female.

scribed first, it claimed priority as the formal species name. Interestingly, there is yet an additional subspecies that is found only in Florida.

This nomenclature-changing event occurred in 1968. Subsequently, when English language names were given, all of the subspecies received the name, Variable Dancer, to recognize the diverse appearance of the species over its range. We do not know the biological significance of the dark- and clear-winged forms. Over time, centuries or, more likely millennia, the different forms may become reproductively separate, in which case they would again have separate species names [89]. Because as full species they cannot interbreed, both could occur together with slightly different habitat preferences or behavioral traits. If and when that situation occurs, this book will be long forgotten and humans may have changed much more.

The Violet Dancer is common near flowing water, particularly in the northern Piedmont region, where it will perch on rocks and logs in and along streams like the Brandywine or White Clay creeks. It also occurs on the Coastal Plain streams like the Choptank and Nanticoke rivers, but is less common there. As indicated by its Latin subspecies name, it is violet or purple. Although other damselflies may have a little purple coloration, this is the only species in our area in which the males are predominantly purple. While Violet Dancers fly from late May through early October, they are most common in June and July.

100 PIEDMONT STREAMS
Argia moesta (Powdered Dancer)

All but the northern fringe of the Delmarva Peninsula is on the Atlantic Coastal Plain that rarely rises more than 70 feet above sea level. The Coastal Plain is composed of layers of sediment. That material was eroded from the ancient Appalachian Mountains, transported by the streams and rivers that flowed into the Delaware and Chesapeake bays, and deposited in the floodplains, sandbars, and deltas that formed the Coastal Plain over millions of years.

The process of erosion continues today with the Susquehanna and Delaware rivers and their many Piedmont tributaries transporting sediment and depositing it in the Chesapeake and Delaware bays. The characteristics of Piedmont streams that continue to erode the continent differ from Coastal Plain streams.

The boundary between the Coastal Plain and the Piedmont occurs a few miles north of the Chesapeake and Delaware Canal where the relatively flat terrain of the Coastal Plain gives way to low hills. The highest point in Delaware is just under 450 feet above sea level near the Pennsylvania state line. The Brandywine Creek, White Clay Creek, Red Clay Creek, Elk Creek, and others streams have carved valleys between these hills. They flow over pebbles, cobbles, and larger rocks with sand and gravel accumulating in eddies and slower sections. This contrasts with Coastal Plain streams that generally flow slowly over beds composed of sand and mud with only a little gravel. While some damselflies occur on both types of streams, others have strong preference for one or the other. The Powdered Dancer only occurs on Piedmont streams and typically perches on rocks in the middle of riffles and short rapids.

The easily identified Powdered Dancer is a fairly common damselfly through the summer in its preferred habitat. The male's light gray color gives it the appearance of being covered with a light coating of talcum powder. The thoraces of females may be blue or brown. When disturbed, Powdered Dancers usually fly to another rock in the stream. Oviposition occurs in plant stems near the surface with the male sticking straight up still clasping the female's prothorax. The female may submerge while ovipositing. Males release their grip on the female before they are fully submerged. If oviposition sites are few and the Powdered Dancers are common, hundreds of tandem pairs may be crowded together in a small area. Powdered Dancers first appear in late May and finally disappear in September.

A recent study of Piedmont streams reveals that these streams look a lot different now compared to before European colonization (*92*). Today these streams have vertical banks sometimes 15 feet high. These banks are composed of sediments deposited behind mill pond dams at a time when clear cut forests and poor agricultural practice led to rapid top soil erosion. Few dams remain and the streams have cut down through the sediment to their old stream beds. However, the character of the streams is not the same as they were centuries ago. It appears that Powdered Dancers have had to put up with human alterations of their home for nearly 300 years.

JIM WHITE

The Powdered Dancer is a common damselfly that perches on rocks and streamside vegetation along swift-flowing Piedmont streams. It can be quite abundant and sometimes congregate densely when oviposition sites are scarce [103].

101 TRIBUTE TO LEONORA K. GLOYD (1902-1993)

Argia sedula (Blue-ringed Dancer)

Female damselflies are notoriously difficult to identify without considerable experience and at least a hand lens. I confess that more often than not, I don't try unless the female is collected together with a male. In that case I assume the male has things figured out. The species-specific characteristics of female damselflies frequently reside in the shape of a small surface on the front of the thorax, right behind the head where the male attaches prior to mating. The males, by contrast, typically have distinctive color patterns on their thorax and abdomen, and the shape of their abdominal appendages that attach to the female are easier to see and differentiate. Consequently, I will identify male damselflies with some confidence, but don't trust my skills with some female damselflies.

In 1974, I collected a female damselfly on Brandywine Creek in northern Delaware that I thought was a Blue-ringed Dancer. But, as I noted above, I wasn't sure. Furthermore, I have not seen the species on the Delmarva Peninsula since, though it certainly could occur here. While leafing through some old correspondence, I discovered a letter written in 1983 from Leonora K. Gloyd, the world expert on Argias at the time. She had examined and confirmed the identity of a male specimen collected September 4, 1944 on the Brandywine

RICHARD ORR

The Blue-ringed Dancer has not been seen for sure on the Delmarva Peninsula since 1944 when it was found on the Brandywine Creek near the Pennsylvania border. This male was photographed along the Potomac River.

Creek by Robert Gibbs. Thus the Blue-ringed Dancer is included in this book based on a single specimen collected decades ago.

Leonora K. Gloyd, known to everyone as Dolly, was born in Kansas on a wheat farm (*110*). She received a Master of Science degree in vertebrate embryology from what is now Kansas State University. Her serious interest in dragonflies and damselflies began in 1929 when she was employed to assist E. B. Williamson in curating his huge collection that had been donated to the University of Michigan's Museum of Zoology. Williamson had hoped to do a monograph on the genus Argia, which has many closely related species. However, he died in 1932 and the monumental task was passed on to Dolly Gloyd who eventually had 75,000 specimens to examine. It was Rosser Garrison who completed the task not so long ago.

Among Argias, the Blue-ringed Dancer is fairly common and widespread. It prefers larger rivers and is known from the Potomac, Susquehanna, and Delaware rivers. While there are related species in the tropics that could cause confusion, male Blue-ringed Dancers are not likely to be confused with other species in the northeastern United States. As its name implies, it has prominent blue rings at the base of each abdominal segment with black in between. Although I have spent many hours wading the Brandywine Creek at different times throughout the summer, I have yet to see a male Blue-ringed Dancer and thus still question the identity of the female I collected many years ago.

102 22ND CENTURY DAMSEL?
Argia tibialis (Blue-tipped Dancer)

Although found throughout most of the eastern United States, the Blue-tipped Dancer is not known from central Pennsylvania or New England, where I acquired and cultivated my interest in dragonflies and damselflies. Consequently, it was new to me when I first saw it after I moved to Delaware. This damselfly with a dark purple thorax and black abdomen, save for its blue tip (males), can be quite common on Coastal Plain streams from June through August. It clearly prefers slow-moving, forested streams with silty bottoms. There, adults like to perch on the barkless trunks of fallen trees, twigs, grass, and bare ground along the shore and in nearby woodlands. Only rarely have I seen it on the Piedmont sections of streams like Elk Creek in Cecil County or White Clay Creek in New Castle County, though it occurs on the slower stretches nearing sea level.

To see a Blue-tipped Dancer, pick a hot, sultry summer day and go to a bridge that crosses a Coastal Plain stream like the Choptank or Pocomoke. More often than not, there will be one or more Blue-tipped Dancers perched on the guard railing or cement surfaces around the bridge. If not there, look for a dark damselfly with its wings folded back against its abdomen on dead branches by the water. Like other dancers, they tend to stay low to the ground. Often they will be the most common damselfly present.

Many of our dragonflies and damselflies display bright colors, live in unusual habitats, have distinctive behavior, or are rare. These features make them attractive and interesting. One could take pity on the poor Blue-tipped Dancer because it is not spectacularly colored, lives in a well-represented habitat, has no particularly unusual behavior, and is quite common. Even most females are drab brown, but a small number have light blue thoraxes with the characteristic forked black shoulder stripe. However, those features may be a blessing when one considers survival into the coming century. My guess is that the Blue-tipped Dancer will still be here after many of our better known species are relegated to the field guides for locally extinct fauna.

The blue phase female of the Blue-tipped Dancer (bottom left) is unusual compared to the typical female shown in tandem with a male in top image. A male is shown at the bottom right.

103 ARGIA—THE NEWS JOURNAL OF THE DRAGONFLY SOCIETY OF THE AMERICAS

Argia translata (Dusky Dancer)

The Dragonfly Society of the Americas (DSA) [84] formed in 1989 at a meeting in Johnson City, Tennessee. (Please note that "dragonfly" here is used in its original sense that applies to all Odonata.) Among its objectives were to promote interchange between amateur and professional odonatologists, support wetlands and habitat conservation as a means of species preservation, sponsor regional and international meetings, document and catalog data on species distribution, and establish a news journal. The news journal was named ARGIA in honor of the damselfly genus *Argia*. It is the only genus of dragonflies or damselflies restricted to and also represented in every country in the Western Hemisphere.

DSA and its news journal have flourished. ARGIA is one of the few periodicals I read cover-to-cover. In contrast to professional journals in which articles have a formal style, articles in ARGIA tend to be conversational, personal, and engaging. They range from brief excerpts from recent e-mail correspondence to entertaining accounts of gringos in Latin American jungles and other places in pursuit of Odonata. Biographical, anthropological, paleontological, and aeronautical articles related to dragonflies and damselflies occur interspersed with expected reports on their behavior and range extensions.

Among the many species of *Argia*, commonly known as dancers, the Dusky Dancer probably has the widest geographical distribution of any of the more than 100 species. It is known from Ontario in Canada to Argentina in South America. While it may have a wide distribution in the Western Hemisphere, its distribution is limited in our area to a few larger, clean-flowing Piedmont streams [113] such as Brandywine Creek in New Castle County, and Elk Creek, Little Elk Creek, and the Susquehanna River in Cecil County. It has recently reappeared on White Clay Creek in Delaware, which is a hopeful sign of the stream's gradual recovery. In all locations it is uncommon. It tends to fly later in the season, from July into September. The later flying season also applies in New England where the Dusky Dancer occurs on large, nutrient-poor ponds such as the famous Walden Pond, where I once found it.

Mature male Dusky Dancers are mostly black with narrow, often interrupted, blue rings on most abdominal segments. Females are brown and black with only pale touches of blue. They perch on rocks and sticks along streams.

The Dusky Dancer occurs locally and uncommonly on Piedmont streams of the Delmarva Peninsula near the Pennsylvania border. A pair of Dusky Dancers (male with blue eyes) ovipositing surrounded by three pairs of ovipositing Powdered Dancers [100].

104 NAME-CALLING
Chromagrion conditum (Aurora Damsel)

Taxonomy is about names and hierarchical relationships [1]. Originally it was simply about classification of organisms based on shared characteristics, not necessarily relatedness. The challenge was to determine meaningful categories that could sort organisms into different groups. Linnaeus, the Swedish biologist who established the current system of classification in the 1700's, did not have the benefit of understanding evolution or a knowledge of genetics provided a century later by Darwin and Mendel, respectively, and biologists after them. Consequently, relatedness, in an evolutionary or genetic sense, was not an underlying principle of Linnaeus' system of classification. However, the genetic continuity implicit in evolution provides a conceptual underpinning for taxonomy. Now many systematists use DNA sequences to establish taxonomic relationships so that taxonomic relationships mirror evolutionary relationships. The hierarchical classification of the Aurora Damsel is given as an example below.

TAXONOMIC LEVEL	SCIENTIFIC NAME	MEANING
Kingdom	Animalia	Animals
Phylum	Arthropoda	Joint-footed animals
Class	Hexapoda	Six-legged animals, (Insects)
Order	Odonata	Dragonflies and Damselflies
Suborder	Zygoptera	Damselflies
Family	Coenagrionidae	Pond Damselflies
Genus	*Chromagrion*	
Species	*conditum*	Aurora Damsel

Looking at the classification of the Aurora Damsel, we see that among animals and within the arthropods, it is an insect because it has six legs. It is not a worm, crab, or spider. Among insects, it belongs to the order Odonata, which includes two major suborders, Anisoptera and Zygoptera, corresponding to dragonflies and damselflies, respectively. These names refer to the difference in the shape of the front and hind wings. In dragonflies the front and hind wings have different shapes with the hind wing broader at the base, while in damselflies the wings have a similar shape. Of course, there are other differences, such as the shape of the head and the way the wings are held at rest. Although there are many families of damselflies, the Coenagrionidae represent more than half of all species. They tend to be small and often are seen around ponds, thus the name Pond Damsels. The Aurora Damsel is classified within this group and is unusual because it has no other relatives in its genus; that is, it belongs to a monotypic genus.

The male Aurora Damsel is a gorgeous blue and yellow damselfly with black markings. Some females are similarly colored while others are black and yellow. It prefers small pools associated with spring-fed streams. Unlike other Pond Damsels, it holds its wings slightly apart, making them appear somewhat like the Spreadwing family of damselflies. However, the two can easily be distinguished in the field because the Aurora Damsel perches with its body horizontal to the ground. Spreadwings perch on vertical stems, and their abdomens hang down. Furthermore, there are no bright blue spreadwing species. Aurora Damsels do not seem to be as common as they were in the 1970s. Their flight season is short, beginning in May and extending into early July.

HAL WHITE

The splash of bright yellow on the thorax distinguishes the Aurora Damsel from bluets. Also, in contrast to other blue damselflies which hold their wings tight together when perched, the Aurora Damsel holds its wings slightly apart.

105 EXTRA EYES

Enallagma aspersum (Azure Bluet)

Going eye-to-eye with a spider using a magnifying glass can be a spooky experience because spiders have multiple eyes, commonly four pairs of simple eyes. They lack the large, multi-faceted compound eyes of insects and crustaceans. When going eye-to-eye with a dragonfly or damselfly, one cannot miss the large compound eyes that dominate the head. What often goes unnoticed are three additional simple eyes that lie near the top and middle of the head. These are perhaps best seen in damselflies, where the compound eyes are widely separated on the dumbbell-shaped heads.

These extra eyes are called ocelli. One is located in the middle in the front, and the other two are laterally separated. Tillyard, the famous Australian dragonfly expert of the last century, concluded that these eyes provide close-up vision over and to the sides of the head while the large compound eyes have much greater range and can detect motion (84). Larval Odonata lack ocelli until they near metamorphosis—another one of those observations that prompt curious biologists to ask "Why"?

In adult damselflies, the markings near the ocelli vary among species. Presumably, they have a role in species recognition. Among bluets, this region of the head can range from mostly black with small postocular spots (spots on the head behind and between the compound eyes) to almost no black and large postocular spots. A species with large postocular spots is the Azure Bluet.

The three ocelli (extra eyes) of this male Azure Bluet can be seen as small reddish dots arranged in a triangle in the middle of the black area between the eyes. The blue patches on the back of the head are called post-ocular spots. The size and shape of these spots is often used in species identification.

This tandem pair of Azure Bluets shows that the appendages at the end of the male's abdomen grasp the front of the females thorax. This is characteristic of damselflies. In contrast, male dragonflies grasp the top of the female's head.

Azure Bluets prefer fishless ponds [92] and frequently inhabit sphagnum-edged ponds. They also regularly appear at Delmarva bays. They skim over the calm water surface. The thorax is mostly blue as are the last three segments of the abdomen with black in between. They fly from late May through August and probably have more than one generation per year.

106 INVASIVE SPECIES?
Enallagma basidens (Double-striped Bluet)

When we talk of invasive species, we normally refer to non-native organisms that humans have introduced to a new location with a favorable environment and a lack of natural predators. As a consequence, the population of the introduced organism explodes and displaces native species. We see this phenomenon with numerous plants like purple loosestrife, dandelion, kudzu, and mile-a-minute weed. The principle also is illustrated by the introduction of rabbits into Australia, goats and rats onto innumerable islands, and Gypsy Moths into the northeastern United States. For that matter, modern humans are probably the prime global example of an invasive species.

As with the much better known birds and plants of Hawaii, the native insects there display striking examples of speciation akin to Darwin's finches in the Galapagos Islands. However, the islands of Hawaii have an unpleasant distinction in the world of invasive dragonflies and damselflies. The trade in aquarium plants to Hawaii has included hitchhiking dragonfly and damselfly larvae that have established themselves on the islands. In only a matter of decades, several nonnative species have become the most common species there. By contrast the highly speciose damselflies endemic to the islands have become exceedingly scarce.

An example of an invasive dragonfly species in the continental United States is the Scarlet Skimmer (*Crocothemis servilla*). It comes from southeast Asia and is now well established in Florida. Its ocean-hopping presence is also likely due to the aquarium trade.

There needs to be a name other than "invasive species" for native species that have greatly expanded their geographic ranges in a short time, not by human introduction though perhaps in part due to human activities. In these cases, natural predators may prevent population explosions as ranges expand. Species show up now in small numbers in places where they were rarely or never seen before. As global warming continues, undoubtedly some southern species will expand their ranges northward and northern species will retreat farther north.

The Double-striped Bluet represents one of the most dramatic range expansions detected in the United States [88]. Before 1930, it was unknown north and east of Texas. In a little over 50 years, the Double-striped Bluet expanded its range to the East Coast and as far north as New York and Connecticut. On Delmarva, this small blue damselfly is not common, thus it differs from the typical invasive species that become abundant and displace native species.

As with other bluets, the dark marking on the head and thorax are similar in the male and female Double-striped Bluet but quite different on the abdomen. The degree to which the female acquires the blue coloration varies as is seen in the two pictures here.

As its name implies, the Double-striped Bluet differs from other bluets by having the normal black stripe on the thorax divided in two by a thin blue stripe or by brown in some females. The presence of Double-striped Bluets at any one place seems unstable. It occurs for a few years and then disappears. My records range from late May through mid-September in places as scattered as large mill ponds, a small concrete pond with a fountain, and slow sections of large streams.

107 TRIBUTE TO HERMANN HAGEN (1817-1893)

Enallagma civile (Familiar Bluet)

In the scientific literature, the last name of the person who first describes a species follows its Latin name. Examination of a list of dragonfly and damselfly species from eastern North America reveals the name Hagen after many of them, for example the Familiar Bluet, *Enallagma civile* (Hagen). In this case, the parentheses around Hagen's name indicate that although he was the first to describe the species, *civile*, he described it in a genus other than *Enallagma*. Someone came along later, studied it and related species, and decided to reclassify it.

Hagen never had seen or collected the bright blue Familiar Bluet alive when he described it in 1861. How could that be? At that time, he was a practicing physician in Königberg, East Prussia, now Kaliningrad, Lithuania, where he was a prominent citizen serving as a member of the school board and vice president of the city council. Apparently, his capture of an undescribed species of dragonfly sparked his early interest in these insects and led to his serious study of them. Although he had never crossed the Atlantic Ocean, he became a leading authority on North American Odonata. Naturalists exploring the fauna here sent their specimens to him for identification or description. For anyone familiar with the way that the living colors of dragonflies or damselflies like the Familiar Bluet fade after death, what Hagen probably received was a brittle brown specimen to describe. It is little wonder that permanent structural features that do not change, rather than variable and ephemeral color patterns, dominate species descriptions, both then and now.

In 1867, Louis Agassiz, the renowned Harvard scientist who had been lured from Switzerland, in turn enticed Hermann Hagen to leave his position in East Prussia to replace Philip Uhler as a professor and curator at Harvard University's Museum of Comparative Zoology (15). Even though Hagen was a "museum" entomologist, one would expect he went into the field after his arrival in Massachusetts and became acquainted with the brightly colored Familiar Bluet he had described. It is probably the most common and widespread North American bluet.

Male Familiar Bluets are in fact mostly blue, but females may be brownish as well as blue, or both. In contrast to other damselflies of their size, they frequently fly just above the surface over open water away from the shores of the ponds they inhabit. Sometimes when few damselflies are among the shore vegetation, hundreds of blue damselflies may be skimming over a pond's surface.

Familiar Bluets seem to challenge the wind because they will be perched like weather vanes on stems out from the shore, even in a good breeze. They fly from mid-May to mid-October and have been reported from all but two counties on the Delmarva Peninsula.

Familiar Bluets feed away from water. Unlike the acrobatic pursuit of prey seen with dragonflies, Familiar Bluets and many other damselflies glean their prey. In this feeding mode, they hover over plants and pluck small insects from leaves instead of catching them in flight (73).

The male Familiar Bluet is indeed bright blue. The very small antennae of damselflies are not much bigger than the hairs that cover their thorax. The female of the tandem pair (right) is ovipositing in vegetation beneath the water surface.

108 A CALVERT SPECIES
Enallagma daeckii (Attenuated Bluet)

Philip P. Calvert would certainly rank among the top five 20th century dragonfly specialists from North America. He was born, reared, and lived his 90-year life (1871-1961) in Philadelphia. His parents nurtured his early interests in natural history, and entomologists at the Academy of Natural Sciences of Philadelphia took over from there. By the time he was seventeen he had a passion for dragonflies and was exchanging specimens and communicating with specialists all over the country (*100*). In 1893 at age 22, a year after graduating from the University of Pennsylvania, he published his "*Catalog of the Odonata (Dragonflies) of the Vicinity of Philadelphia, with an Introduction to this Group of Insects*" (*16*). This work includes some of the first references to dragonflies and damselflies from the Delmarva Peninsula. (Note that in 1893 usage, damselflies were considered as a subgroup among dragonflies rather than as a separate group equal to dragonflies as treated in this book.).

Calvert's *Catalog* was the first real treatise on dragonflies and damselflies in North America. Because specialists from Europe had already described most of the species, it was time for someone to gather what was known, generate identification keys, and determine geographic and seasonal distribu-

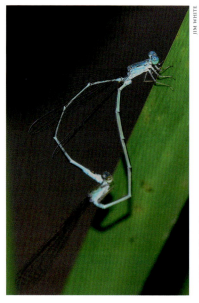

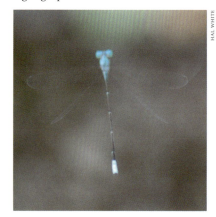

A close look at the female of this mating pair of Attenuated Bluets reveals a couple of water mites attached to the underside of her abdomen near the ovipositor. Note the distinctive dark bands across the female's blue eyes. Above is a male in its slow-moving hovering flight.

tions. In addition to his scholarly coverage of the literature of the time and his skill as an illustrator, Calvert's field work included bicycle expeditions to the Pocono Mountains of Pennsylvania and the Pine Barrens of New Jersey. Thus, in contrast to the specialists who described species they had never seen alive and who had limited knowledge of the location or habitat of origin, Calvert was a field entomologist who observed the various species in their native habitat. Reading Calvert's *Catalog*, one would think that all of the species in this area had been described. However, the list of species now known from the area includes additional species, several with the name Calvert as the describer. One of these species is the Attenuated Bluet, which he described from New Jersey ten years after his book was published (*17*).

The Attenuated Bluet has an ethereal presence. It is larger, more slender, and has a lighter blue color than other blue damselflies, which caused Calvert to describe it as a new genus—a distinction that did not last. When it flies undisturbed, it hovers a lot and glides slowly among dense vegetation at the edges of ponds and swamps in June and July. Although not common, it occurs locally on the Coastal Plain as far north as Massachusetts. This species is known from the Primehook National Wildlife Refuge and is probably more common on the southern half of the Delmarva Peninsula than has been reported. In the hand, its eyes are distinctive, with two dark horizontal bands across them.

109 BREATHING UNDER WATER
Enallagma divagans (Turquoise Bluet)

Growing up, my friends and I would challenge each other to see how long we could hold our breath underwater. Our record for that swimming-hole challenge in the stream in my back yard topped out at about two minutes. We knew that fish had gills and could live underwater, but we were dependent on air. We watched the tadpoles we had brought to school metamorphose into toads and frogs. The teacher told us that tadpoles could breathe under water because they were able to absorb oxygen from the water through their tail and skin. She also noted that although frogs and toads had lungs and could breathe air, they were still able to absorb oxygen from water through their skin.

Good teachers always hope that their students can transfer their knowledge and understanding to new situations. This lesson had the potential to provoke lots of curiosity. However, only after I had been told about how dragonflies and damselflies deal with metamorphosis from an aquatic larva to an air-breathing adult did I generalize the problem and think more systematically about how aerobic organisms live underwater and the challenges that represents as a function of temperature, pollution, and water-mixing.

The larvae of dragonflies have gill-like structures inside their abdomen [63]. They in essence breathe by circulating water in and out of their abdomen. This has the added feature that rapid expulsion of water serves as a form of jet propulsion when they need to escape from predators. While damselfly larvae may obtain a little oxygen through the abdomen, they have a different adaptation. At the end of the abdomen are three caudal lamellae [95], also known as gills. These fan-like often flat structures absorb oxygen from the water. They too provide a means of rapid escape. By undulating their body side to side like a fish, the gills serve as a swimming fin.

The caudal lamellae of damselfly larvae often have species-specific characteristics and thus provide a way to identify larvae. They differ in shape, markings, and pattern of the underlying tracheae that serve to distribute the oxygen to the rest of the body. On the Delmarva Peninsula in June and early July, children around a stream swimming hole quite likely would see the adult Turquoise Bluet. This slender damselfly has a blue thorax striped with black and two entirely blue segments near the tip of its abdomen with black in between and at the very end. At other times of the year when its larva lives underwater and is not turquoise, the larva can be identified by the distinctive pattern and lance-like shape of its caudal lamellae.

The flight season of Turquoise Bluets is relatively short—less than two months. They appear first in mid-May and usually disappear by early to mid-July. Males and females are similar in appearance. They fly along the shoreline and perch within a foot of the water surface on emergent vegetation. As shown, adults can survive under water when ovipositing. Presumably the hairs on their thorax help create a bubble of air that enables them to breathe underwater.

Turquoise Bluets perch on emergent vegetation in and along Coastal Plain streams. As shown above, oviposition can occur underwater. In this case, the male released and flew off while the female moved farther down the stem, remained submerged, and continued ovipositing for many minutes. The larvae have three caudal lamellae, flat fin-like structures at the end of the abdomen, that are used as underwater gills and for propulsion in swimming.

110 BRIDGING THE GAP
Enallagma doubledayi (Atlantic Bluet)

Cumberland County, New Jersey, is just across the Delaware Bay from Delaware. There the Atlantic Bluet is described as locally abundant (*2*). Next door in Cape May County it is considered common and flies from mid-May to mid-October (*78*). In both counties, it occurs at many ponds. It is also known from parts of Maryland and south to Florida, where it flies year round, and on west to Mississippi. Yet until 2007, there were no records of the Atlantic Bluet from anywhere on the Delmarva Peninsula—a rather significant gap in its known distribution. Clearly it had been missed despite the fact that I had searched for years, mostly in northern Delaware. It seemed that one just needed to find the right habitat.

In New Jersey, the Atlantic Bluet is most common at fishless ponds in gravel pits and sandy-bottomed depressions excavated to create the road bed for the Garden State Parkway. Such habitats exist on this side of the Delaware Bay, but they are less frequent, often on private property, and thus less accessible. An additional factor contributing to the apparent absence is that the Atlantic Bluet looks very much like the Familiar Bluet [**107**], perhaps the most common of all the blue-colored bluets. Thus, without actually catching the damselfly and examining the male appendages (females are virtually indistinguishable), one can miss the Atlantic Bluet. Interestingly, the two species rarely occur together, which raises the possibility that speciation in this case was associated with habitat specialization that limited potential interbreeding.

In July 2007, Hans Holbrook with Jim Brighton made the long-expected discovery of the Atlantic Bluet in Wicomico County, Maryland. I suspect that exploration of the appropriate habitats in the sandy areas of southern Delaware and Maryland will turn up many more records. A field identification that is not always reliable is that the blue (postocular) spots behind the eyes are connected by a bar in the Atlantic Bluet, but not in the Familiar Bluet. The male appendages of the Familiar Bluet are longer at the top whereas the appendages of the male Atlantic Bluet are symmetrically rounded and blunt.

BILL HUBICK

JIM BRIGHTON

The Atlantic Bluet undoubtedly escaped detection for many years due to its similarity to other bluets. It is common in southern New Jersey and should turn up in other places on the southern Delmarva Peninsula.

111 AGRICULTURAL RUNOFF AND HABITAT DEGRADATION

Enallagma dubium (Burgundy Bluet)

Despite its dark-reddish purple color, which is spectacular and distinctive, the small Burgundy Bluet is easily overlooked as it perches on lily pads away from the shore. In 1973, Minter J. Westfall, a Floridian and one of the authors of the scholarly volumes, *Dragonflies of North America* (57) and *Damselflies of North America* (95), found this damselfly at Records Pond in Laurel, Delaware. I had never seen it before. Embarrassingly, he spotted it almost under my nose, demonstrating once again that it is hard to see something when you do not know what you are looking for.

At the time, Records Pond was the northernmost known location in the United States for the Burgundy Bluet. A well-established population of this southeastern species was found there and four miles upstream at Trussum Pond during the next two years, but it hasn't been recorded since then despite several searches there in the intervening years. While difficult to prove, its disappearance may well be due to increased nutrients from agricultural runoff that has drastically changed the character of ponds around Laurel and elsewhere.

RICK CHEICANTE

Burgundy Bluets are rarely seen on the Delmarva Peninsula. Look for them perched on a lily pad out from the shore of unpolluted ponds.

The shallow ponds in and near Trap Pond State Park with bottoms of sand and silt used to be covered with beautiful water lilies and a variety of emergent vegetation. Clouds and the blue sky reflected off their surfaces. Now tremendous algal blooms fill the waters each year, and harvesters have been used to remove the algae. Swimming beaches that once overflowed with children are now closed and considered polluted. Trussum Pond, where bald cypress trees give the feel of southern Georgia, is covered completely with a green carpet of duckweed rather than lily pads in the summer. First-time visitors may delight at the scene, but I weep knowing what it used to be.

The impact seems primarily hard on damselflies like the Burgundy Bluet, and not on dragonflies. In a survey of Trap Pond in 2002, I saw only 15 individuals of five species of damselflies in three hours whereas a similar survey in 1975 found 15 species of damselflies, many in large numbers. Clearly, many damselfly species in addition to the Burgundy Bluet have suffered major declines in recent decades. What is going on? If we knew for sure, we might be able to reverse the changes.

Though it is at the northern end of its range and uncommon here, the Burgundy Bluet has a distinction. It is one of only two species of dragonfly or damselfly [115] first discovered and described from the Delmarva Peninsula. Francis Root (45) discovered it at a lily pond in a cypress swamp near Salisbury (69), Maryland. In 2009, after many years of searching, the Burgundy Bluet was rediscovered by Hans Holbrook at a long-abandoned sand borrow pit farther north in Caroline County, Maryland.

112 INLAND SEA
Enallagma durum (Big Bluet)

Have you noticed the vegetation along highways? Certain plants are able to grow almost next to the pavement. This could be due to any number of factors associated with roads, their construction, or chemistry. I remember driving north one morning mile after mile on Interstate 91 in Vermont and trying to figure out why the narrow strip of blooming blue chicory grew right next to the roadway. Was the soil more disturbed there? Was there more moisture from rain runoff? Was it warmer? Perhaps it was due to road salt left over from winter. I suspect someone who knows about chicory could tell me the critical factors. However, it was the possibility of salt tolerance that kept me thinking and wondering.

A newspaper article I once read mentioned how all the pine trees along well-traveled roads in New England were dying from the road salt used to melt snow and ice. I could see the dead and dying pines. The problem is not confined to traditionally snowy country. A recent study near Baltimore showed that runoff from road salt temporarily increased the salinity of some streams to a quarter of that of sea water (*48*). Of course these events are temporary, but consider when these streams flow into ponds where the salt can persist for awhile. How does that affect the pond life? Do salt-tolerant species colonize or increase in abundance and salt-intolerant species disappear? My observations over thirty years at Lums Pond in Delaware make me think this could be happening, although chloride concentrations recorded by the Delaware Department of Natural Resources and Environmental Control between 1997 and 2007 have not shown an upward trend.

The Big Bluet is a robust and handsome species that is known for its salt tolerance. It is common in freshwater estuaries along the northeastern Atlantic coast. It is common along the lower Hudson River, in New York, and at Perryville, Maryland, where the Susquehanna River enters the Chesapeake Bay. Although Lums Pond is near the Chesapeake & Delaware Canal, it is a freshwater pond. However, in recent years some species that used to be fairly common there like the Elegant Spreadwing [93], have become rare or no longer present. At the same time, the Big Bluet, hardly known before 1990, has become the most frequently seen blue-colored bluet there. In this watershed, housing developments and associated roads have covered acres and acres of farmland in a couple of decades. It would be interesting to determine whether the salinity of this freshwater pond has risen significantly over the same time period or whether short term salinity increases missed by periodic sampling might explain the changes in species abundance.

Because the Big Bluet is salt-tolerant and dragonflies and damselflies are considered freshwater insects, Big Bluets are often found where naturalists don't normally look for dragonflies—like the mouth of the Susquehanna River. Once one knows where to look, the Big Bluet turns out to be more common than expected. For example, I discovered the first population in Maine by specifically looking for it in a freshwater tidal marsh. As their common name implies, the Big Bluet is a relatively large, mostly-blue species, with prominent, pointed black abdominal markings. Females are blue or olive green with significantly more black marking. Their flight season extends from the second week of June through the end of September.

The male Big Bluet is quite colorful with black markings contrasting with rich blue on the thorax and abdomen. The female has more black with light green on the sides of the thorax.

113 WHAT MAKES A HOME?

Enallagma exsulans (Stream Bluet)

What enables delicate damselfly larvae to live in a swift-flowing, rocky Piedmont stream? How are they different from the larvae of related species that live in other habitats? What is it about these streams that favors one species over another? These questions, with appropriate word substitutions, apply to every organism and habitat. Adaptations that provide wide or narrow habitat tolerances have evolved in every organism.

Each habitat is unique at some level. Consequently, one frequently asks why a certain species is common at one place and rare or absent at a seemingly

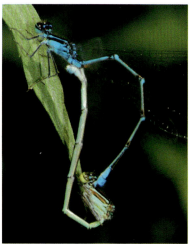

Above, a male Stream Bluet is having a meal. At the left, a pair of mating Stream Bluets perches on stream-side vegetation. The end of the female's abdomen swings around under the male and attaches at the base of the male's abdomen. The thoracic stripe on the female has a characteristic bronze coloration.

similar habitat not far away. Is there a difference in the water due to the soils and local geology? Does the orientation of the habitat cause it to be warmer or cooler? Are there unseen pollutants or particular nutrients present? Are there distinctive microhabitats? Does the presence or absence of certain other species matter? The questions go on and on, and few have answers.

The Stream Bluet is a delicate damselfly that prefers Piedmont streams, while most of its bluet relatives prefer ponds. This implies that there is something about streams that Stream Bluets can tolerate that the others cannot, or vice versa. Because permanent streams on the Piedmont flow faster than streams on the Coastal Plain, their waters are well-mixed and well-oxygenated. They sweep away small particles so that mud and sand accumulate in the slow stretches. The range of temperatures is narrower.

Any organism that thrives in a stream environment must be able to survive floods by finding shelter to avoid being swept away. When waters recede, any aquatic organism left on the floodplain must find its way back to the stream or die. Significant variations in year-to-year abundance can be correlated with major winter floods that scour stream beds. Summer floods may have less effect if the adults are still flying and can repopulate a stream.

The Stream Bluet possesses the characteristics necessary for survival on Piedmont streams because it is common there—a trait shared with the Powdered Dancer [100]. Occasionally, a Stream Bluet turns up on a Coastal Plain stream where there is significant current. Elsewhere, it occurs along the shores of lakes where wave action provides some of the characteristics of a flowing stream. Dates for the Stream Bluet in Delmarva range from early June to the beginning of October. The males are slender and black with blue, a mostly blue thorax, and blue at the base of each abdominal segment except for the next to last segment, which is entirely blue. Females are dark with blue-green markings and a brown shoulder stripe margined by black.

114 LITTLE BLUE DAMSELFLIES FOLLOW THE ICE

Enallagma geminatum (Skimming Bluet)

With the exception of a small strip along its northern edge, the Delmarva Peninsula is part of the Atlantic Coastal Plain. Land elevations rarely exceed 70 feet, and soils consist mostly of silt, clay, sand, and gravel washed down from the Appalachians. It's hard to imagine that 15,000 years ago, a mere blink in geological time, this was a barren tundra with permafrost where mammoths roamed. Their teeth and bones remain as evidence. Sand wedges found in borrow pits attest to bygone permafrost. Strong winds off the continental glacier blew sand and dust around, carving depressions in the landscape. This is the current best explanation for the origin of the many wetlands known as Delmarva bays. Sea level was so low that the Chesapeake Bay and Delaware Bay did not exist. Major rivers flowed in their place on and over what is now the submerged continental shelf. Few, if any, dragonflies or damselflies lived here then. Our entire fauna colonized this area from the south. Rapid climate and habitat changes favored adaptable species. Enter the bluets.

Mark McPeek, a biology professor at Dartmouth College in New Hampshire, and his students have studied bluets using molecular techniques to determine their relationship to one another (*14*). He discovered that mitochondrial DNA was almost identical among many species. Compared to other types of DNA, mitochondrial DNA evolves rapidly. I, wearing my evolutionary biochemist's hat and being familiar with the distinctive species of bluet, found the result quite unbelievable at first. It implied that these many species evolved from a few since the end of the last glaciation! For comparison, humans and chimpanzees are closely related, but based on similar studies, their last common ancestor lived more than 5 million years ago. Subsequent research by McPeek has supported the early evidence, and I accept the remarkable conclusion that the bluets represent a recently evolving species swarm (a group of closely-related species).

Often a species swarm contains species that are very similar or even seemingly identical. Not so with the Skimming Bluet, a member of this species swarm and a wide-spread resident of Delmarva. It is easily distinguished from very close relatives by its relatively small size and distinctive blue patterns on both males and females. In males the abdomen is mostly black except for the last two segments, which are blue. Females have paired blue spots on the top of abdominal segment 8. They characteristically fly low over the water and perch on floating vegetation. They first appear in May and occasionally persist into September.

The Skimming Bluet is one of the smallest bluets. It usually perches on floating vegetation such as lily pads.

115 TYPE LOCALITIES AND TYPE SPECIMENS

Enallagma pallidum (Pale Bluet)

Although not far from major East Coast cities, the Delmarva Peninsula was off the beaten path until bridges were constructed to the Eastern Shore of Maryland and from Virginia. Collectors would have need to plan a trip to Delmarva because few would pass through on the way to somewhere else and stop at a roadside stream or pond to net a new state record. Consequently, there are few specimens or published reports on dragonflies and damselflies from the region from the past two centuries.

As sometimes happens, gems are waiting for those who explore a neglected area. Francis Metcalf Root, a mosquito specialist and a parasitologist on the faculty of Johns Hopkins School of Hygiene and Public Health in Baltimore (45), must have had uncommon luck. He discovered not one, but two undescribed damselfly species within a year near Salisbury, Maryland, while studying breeding places for a type of disease-carrying mosquito. The Pale Bluet (68) and the Burgundy Bluet [111] (69) that Root described are the only species for which the type localities are on the Delmarva Peninsula.

When a new species is described, its characteristics are based on a limited number of specimens. These are the *type specimens* and they were collected at

The Pale Bluet is rare on the Delmarva Peninsula. This male was photographed by Hans Holbrook after he had collected it from a perch on dead branch near the water surface at a pond in Caroline Co. Maryland, the northernmost known location for this species.

the *type locality*. In the lingo of taxonomists there are different kinds of types, e.g. holotype (the specimen selected for formally describing a species), paratypes (additional specimens included in the original species description), allotype (specimen of the opposite sex from the type specimen), topotypes (specimens collected at the same locality as the holotype), and so on.

Root described the type specimens of the Pale Bluet from a pair and a male he collected on July 26, 1922, probably at Johnson Pond on the Wicomico River in Salisbury. The type specimens are deposited in the Academy of Natural Sciences in Philadelphia. As its common name implies, the color of this bluet is pale blue, but in the field it would be hard to distinguish from several other bluet species that have blue thoraxes and blue-tipped abdomens with intervening dark segments.

While this little-known species has been found on the Coastal Plain to the south, the range of the Pale Bluet extends northward only a few miles into Delaware, where it was found in the vegetation below the spillways at Trussum and Trap ponds in the 1970's. Unfortunately, it has not been found there in recent years, having apparently suffered the fate of other species including the Burgundy Bluet [111] that he described. Fortunately, the Pale Bluet still occurs around Salisbury, Maryland, its type locality, and was recently discovered in the Idylwild Wildlife Management area of Caroline County, Maryland, now the northernmost known population.

116 FLY UNITED
Enallagma signatum (Orange Bluet)

Long before United Airlines' catchy advertising slogan, male and female drag-onflies and damselflies flew united. This behavior is called flying in tandem. The dragonfly couple is connected head-to-tail with the male always in front. The appendages at the tip of his abdomen grasp the top of the female's head, in the case of dragonflies, or the front of the thorax for damselflies. Because form-ing this connection is a prelude to mating, the process has evolved to avoid mistaken identity. As a consequence, the physical points of contact between the sexes are often structurally distinct for each species. This also means that the male appendages are particularly useful to taxonomists in distinguishing closely-related species.

When a female flies into a male's territory, he will appear to attack her as he tries to make the tandem connection. If successful, the pair typically flies away to a tree branch or other perch, where mating takes place. A female can mate with different males, but each male can remove the sperm package of previous matings. Thus, from the perspective of siring offspring, it is important to be the final suitor before egg-laying occurs.

This logic seems to explain the various guarding behaviors that males have towards ovipositing females with whom they have mated. The ultimate guard-ing is to remain in tandem after mating and during oviposition. This physical connection assures that no other male can mate with the female. Some species even fly in tandem for long distances during migrations. That way, when they get to their destination, they do not have to look for mates that might otherwise be rather scarce.

Damselflies in the bluet family often fly united for many hours. As a conse-quence, a large proportion of the observed individuals are paired up. This is the case for the oxymoronic Orange Bluet. This common, slender species prefers flying in the shade and under vegetation along the edges of ponds and slow-moving streams. Though not identically colored, both males and females are brownish orange with dark markings, which is distinctive. Orange Bluets fly from late April through the summer until early October.

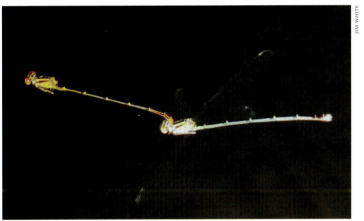

Orange Bluets often prefer the late afternoon shade along the shore of ponds and slow-moving streams where they perch on twigs in the water. There, tandem pairs often fly united. Note how the legs are pulled in close to the body during flight.

117 NOTICING DIFFERENCES
Enallagma traviatum (Slender Bluet)

The human mind seems programmed to notice differences. I have traveled to and from work by the same route for several decades, yet I would be hard pressed to describe in any detail the familiar houses or landscaping that I pass almost every day. However, if a tree comes down, a house is painted a different color, or there is a new car in the driveway, I notice. I guess natural selection made sure changes and difference deserved our attention.

In science, the anomalies attract attention, provoke curiosity, and often lead to further study and new understanding. Who cares about the routine or the common place? The changes and differences raise questions and inspire inquiry.

Along the East Coast, the Slender Bluet is sometimes a common resident of permanent ponds, where it perches in the pond-side vegetation. Abundance seems to vary from year to year. Both males and females are relatively delicate, hence their common name. While there are other species that might be confused with the Slender Bluet, a close look at the form of the male appendages reveals that they are diagnostic. Once one knows the habitat, behavior, and appearance of the Slender bluet, it can be identified "on the wing" without catching it and examining its appendages. That's when it becomes routine, and we no longer pay close attention.

A number of years ago, T. W. Donnelly found some damselflies in east Texas that looked different enough to call them a new species (27). Later examination of specimens of the Slender Bluet from all over the eastern United States showed that what he had discovered was a subspecies of the Slender Bluet that occurs west of the Appalachian Mountains. Knowing that there are two forms raises questions. Where exactly are the boundaries between the two forms? Are there places where the two forms are in contact, but do not hybridize? Do the two forms have different habitat preferences or distinguishing behaviors? None of these questions would have meaning without the recognition that two forms exist. The comparative approach in biology underlies much of what we know. It serves as a lesson. Be observant and don't dismiss the familiar, because it may actually be different and more interesting than you assume.

Look for Slender Bluets in June and July in the vegetation along the shores of Delmarva ponds. Both males and females have blue-tipped abdomens and often fly in tandem. Each of the mostly black segments has some blue at its base.

This tandem pair of the Slender Bluet is perched in the vegetation along the shore of a larger pond.

118 WHEN EVENING SHADOWS START TO FALL

Enallagma vesperum (Vesper Bluet)

Although 1919 seems quite long ago, it is still fairly recent for the first description of a dragonfly or damselfly from eastern North America. Ninety percent of species found on Delmarva had been described by then. Considering that the Vesper Bluet is reasonably common, why did it evade discovery for so long? Undoubtedly, its behavior, or rather the behavior of entomologists then (and now), provides an explanation.

In my normal visits to wetlands, I rarely see a Vesper Bluet. Years go by without my seeing this beautiful species with a bright yellow thorax. That is because I am in the field during the day and want to get home for dinner. I use the evening to record my observations and curate any voucher specimens I have taken. I suspect I would see many more if I changed my ways.

In early July of 1979, I camped with my family at Trap Pond State Park in southern Delaware. I could eat dinner by the water and enjoy the sun setting over the water. As usual, I had spent much of the day wading the shoreline, exploring the streams, and walking the woodland roads looking for and finding an impressive number of species of dragonflies and damselflies. The Vesper Bluet was not among them. Then about 8:00 p.m., while looking out over the water after the wind had calmed, I noticed a yellow damselfly perched on a lily pad near the shore. A closer scan of the area revealed more Vesper Bluets perched on lily pads. I have no idea where they spent the day, but they were fairly common near dusk. I have had similar experiences in Canada and New England.

If the Vesper Bluet didn't have such a distinctive activity pattern, it probably would be called the Yellow Bluet. The only part of its body that is blue is the ninth, or next to last, abdominal segment. Otherwise, the abdomen is mostly black, and the thorax is lemon yellow. The thorax in females is yellow, yellow green, or pale blue.

I haven't revisited Trap Pond in the evening, so I don't know whether the Vesper Bluet is still common there. However, I have noted a startling decline in the diversity and abundance of many damselfly species there, possibly due to changes in water quality. This pond once was covered with lily pads [111], but now has few, and it experiences massive algal blooms each year from agricultural runoff. This ominous change leads me to worry that Vesper Bluets would be hard to find there now.

The camera flash generates a shadow that gives the mistaken impression that this male Vesper Bluet is in bright sunlight. Actually, it is in the shade near dusk.

119 BLACKENED BLUET

Enallagma weewa (Blackwater Bluet)

As a resident of northern Delaware with a day-time job, many of my ventures into the field tend to be nearby and north of the C&D Canal. Trips farther afield on the Delmarva Peninsula, though not far by standards elsewhere, don't happen on the spur of the moment and thus occur less frequently. This bias has implications. I may not see for several years species that occur regularly farther south on the Coastal Plain or near the shore. It also gives the mistaken impression that the dragonfly and damselfly faunas of New Castle and Cecil counties are richer than those of other counties. I'm sure that if I lived in Seaford or Salisbury, I would happily stay south of the C&D Canal and some species that I now consider uncommon would turn out to be frequent and widespread.

Hoping to redress my geographic sampling bias, I began to take more trips south starting in April, with predictable results. On a visit in mid-June to the Choptank River near Goldsboro, Maryland, I encountered several tandem pairs of a slender, mostly black damselflies perched or hovering in the streamside grass where the water was fairly swift. At first I had no idea what the species was, but soon recognized it as the Blackwater Bluet, a species that I had not seen in over 25 years. This southeastern U. S. species occurs as far north as Rhode Island, but it has never been seen on Delmarva north of the C&D Canal. Clearly, it prefers the slower streams of the Coastal Plain. Further south, these streams

JIM WHITE

This shade-loving pair of Blackwater Bluets is perched in the vegetation along the banks close to the water of the Choptank River in Maryland.

are often peat-stained, and thus the name, Blackwater Bluet. Given its color, it could just as easily have been named the Blackened Bluet to go along with the several Bluet species that are not blue.

So often one sees what one expects to see. Without the proper "search image", it is easy to overlook things in plain sight such a Blackwater Bluet when it is an unfamiliar species in a microhabitat not usually explored. When a group of specialists spends a day together in the field, they frequently see different things. Their unique past experiences have prepared them to recognize distinctive niches where they look for associated species that others don't expect.

Having become reacquainted with the Blackwater Bluet and its habitat, I soon started looking for it and finding it along other Coastal Plain streams where there seemed to be a little more current. From now on, when exploring Coastal Plain streams during the summer, I will examine the streamside vegetation within a foot or so of the water and look for slender black damselflies. Both sexes have a small blue spot near the tip of their abdomens. The female has brown and blue stripes on her thorax while the top of the thorax is black in males.

120 FEMALE PARENTS WITHOUT PARTNERS

Ischnura hastata (Citrine Forktail)

In the contemporary world of science, articles about insects and natural history appear infrequently in the premier science journals. Thus, heads turned not so long ago, when an article about damselflies in the Azores appeared in the prestigious journal *Nature*.

Oceanic islands have fascinated evolutionists ever since Darwin's infatuation with the Galapagos. Although such islands support few dragonflies and damselflies because colonization events and permanent freshwater habitats are rare, those species that do occur are often interesting. Perhaps a single fertile female founds an island population that subsequently expands without competition. Over time, separate populations may adapt to different empty habitats and eventually evolve into distinctive endemic species. This has happened in the Hawaiian Islands, where species of the endemic genus of damselflies, *Megalagrion*, evolved. Unfortunately, they are now threatened by human activities and the accidental introduction of several continental "weed" species [106].

Surveyors of the limited diversity of dragonflies and damselflies in the Azores were puzzled by the absence of male Citrine Forktails (74). Not a single male was found among over 300 females collected. Typically, males of any species are more common in collections than females [2]. Thus the apparent absence of males was not a case of overlooking the bright yellow, but very small, males. Rearing of larvae collected on the islands produced only females. They in turn produced fertile eggs in the absence of males. A succession of nine, maleless generations was produced in the laboratory (67). So far this is the only documented instance of parthenogenesis in dragonflies or damselflies.

The Citrine Forktail is found across the Eastern United States and southwest to California. It is known from every county on the Delmarva Peninsula. In contrast to the Azores, we can also see the brightly-colored males. I often ponder when and how this tiny damselfly made it from North America 2500 miles over the open ocean to the Azores when all the other species there originated from Europe or Africa. Some day genetic studies may resolve those questions.

At a little over three-quarters of an inch, the Citrine Forktail is rivaled only by the Sprites [126, 127] as our smallest damselfly. They fly among dense emergent vegetation, often sedges, growing in shallow warm water along pond shores and seepage areas and may be locally abundant. They fly from early May well into November. While the striking yellow abdomen and the green and black striped thorax of males are distinctive, mature females are harder to identify. They are black and gray and resemble females of other forktail species. Furthermore, the young females are orange, but with less black on the abdomen than the orange female of the Eastern Forktail. Although one need not examine wing venation for identification, the unique forewings of the males have pale red stigmas (the dark spots near the front tip of wings) that do not touch the edge of the wing.

The tiny but bright yellow male Citrine Forktail occurs abundantly from May to October in sedge meadows. The male above is attempting to mate with a gray form female. The male at the right landed on a hungry spider.

121 FEMALES OF DIFFERENT COLORS
Ischnura kellicotti (Lilypad Forktail)

For many species of dragonflies and damselflies, the female looks different from the male—a phenomenon called sexual dimorphism [75]. If that were not enough to confuse the beginner trying to identify species, in some species, particularly among the forktails, females come in two or more color forms and those color forms may change during maturation. A species that exemplifies this strikingly is the Lilypad Forktail.

Female Lilypad Forkails come in three quite different color forms. One has bright blue markings and looks similar to the male. Another has bright, porcelain orange markings. And a third type develops a pruinosity that gives it a gray color, which some think comes with aging of the other two forms. All occur together. This polychromism has puzzled biologists because the selection that maintains these forms over wide geographical ranges, and in other species as well, is not obvious. The males that carry genes for these color forms are all bright blue. That males have blue markings on the thorax makes them a bit unusual because they lack the green markings of other local forktails. They look more like bluets and might be confused with Skimming Bluets [114].

As the common name implies, Lilypad Forktails often perch on lily pads. While other damselflies have the same habit, none are as stout, and none have the distinctive, slightly arched abdomen characteristic of the Lilypad Forktail. The more delicate damselflies may lift off and land gently with a hovering flight. By contrast, the Lilypad Forktail seems almost to jump up and plop down on lily pads. Because lily pads tend to grow in deeper water, it is often difficult to approach adult Lilypad Forktails. Binoculars certainly help in spotting them. Oviposition occurs in lily pads, and the larvae are associated with the underwater parts of the lily pads.

Perhaps because lily pads grow away from shore, records of the Lilypad Forktail are scarce. I have found it regularly in a long-abandoned and flooded sand borrow pit in Kent County, Delaware, and Caroline County, Maryland. It occurs in several lily ponds on Virginia's eastern shore. Both males and females are conspicuous there on lilypads from early May into September. They are known along the East Coast as far north as southern Maine.

The bright, porcelain orange female of the Lilypad Forktail (top) looks quite different from the blue males (bottom). Some females do look like the male and as they age develop the gray pruinescence of the female in the middle.

122 TINY EXCLAMATION MARK

Ischnura posita (Fragile Forktail)

At less than an inch, the tiny and inconspicuous Fragile Forktail frequently goes unseen. However, it is certainly our most common and widespread damselfly. Fragile Forktails start emerging as early as the first week of April and are still around as late as mid-October or early November. They have wide habitat tolerance and thus occur at virtually every pond, marsh, or slow-moving stream and at other places as well. Often they fly among the stems of emergent vegetation and, because the male's mostly black abdomen is hardly visible, they appear only as small green dots moving from perch to perch. A closer look reveals that the color comes from an interrupted stripe which is shaped like an exclamation point on the top of the thorax just in front of the wings.

As is true for other forktail species, the females come in strikingly different color forms [121]. The most common form is a uniform light gray. Young females have light blue markings that become gray with age. One wonders whether this polymorphism has some significance in mate recognition that we haven't figured out yet.

In most species of Odonata, males are seen much more frequently than females, There are times, however, when female Fragile Forktails seem to be much more common than males and tend to congregate around floating vegetation. Males should have no difficulty finding a mate then, but where are they? The answer awaits a curious determined and perhaps lucky naturalist.

Observers soon learn that dragonflies and damselflies are hard to find on overcast and rainy days. Given their abundance at certain aquatic habitats, one wonders where they all go when the clouds come. Most certainly they move away from water, so they become more dispersed, and they usually perch, which makes them inconspicuous. The Fragile Forktail seems less affected by clouds. Exploration of the thick, grassy vegetation near their normal habitats on cloudy, or even rainy, days usually turns up a few.

The common name Fragile Forktail was chosen to indicate a small and delicate appearance, but in terms of numbers and habitat tolerance, this is a robust forktail.

The Fragile Forktail, among our smallest damsel-flies, is perhaps the most common damselfly on the Delmarva Peninsula. Males are mostly black with a green exclamation mark on the top of their thorax. Most females are gray.

123 CATCH ME IF YOU CAN
Ischnura prognata (Furtive Forktail)

I have seen a living Furtive Forktail only once, when John Michalski collected this southeastern species unexpectedly in Westchester County, New York. Ever since then the Furtive Forktail has held a particular fascination for me. I noted other isolated records from the Virginia portion of the Delmarva Peninsula, Cape Cod, southwestern Connecticut, and southeastern Pennsylvania. They are spread over many years in June and July. While these occurrences may represent vagrants, the weak-flying, secretive nature of this species suggests that it is an elusive resident. Certainly, I expected it elsewhere on the Delmarva Coastal Plain north of Virginia (*81*). Nevertheless, I searched in vain for it many times in habitats where it might be found. It certainly was furtive.

When one has experience with particular species in the field, a sense of appropriate habitats evolves to the point that one can predict reasonably well which species to expect and not to expect just by inspecting a habitat from the car. Those who know the Furtive Forktail have told me to search for a slender damselfly in the shaded vegetation around swampy ponds and along ditches with stands of lizard tail. I have been trying to develop a search image for something I may never see.

In the summer of 2007, others succeeded where I have not. Jim Stasz, Jim Brighton, and friends discovered the Furtive Forktail in three different counties on the Maryland Eastern Shore. Clearly, they developed a sense of where to look and exploited it. These were the first records for Maryland. Undoubtedly, the first records for Delaware will follow with a little effort. The identity of the quarry will not be in doubt because the Furtive Forktail has some distinctive features. In addition to being longer than our other forktails, rather slender, and mostly black, males have a prominent dorsal projection at the end of the abdomen. The stigma [61] on each front wing is fringed with white. It has narrow green stripes on its thorax and only segment 9 of the abdomen is blue. Young females are orange and become darker on the top of the thorax and pale green below as they mature. The top of the abdomen is dark in males and females.

This first state record for the Furtive Forktail in Worcester Co., Maryland, was photographed by Hans Holbrook in 2007.

124 MATING WHEEL
Ischnura ramburii (Rambur's Forktail)

As noted before, identification of male dragonflies and damselflies often is based on the species-specific shape of the appendages at the tip of the abdomen [22]. The male uses these to grasp the female during mating. In dragonflies, they clamp over the top of the female's head. In damselflies, they clamp onto the front end of the thorax immediately behind the female's head. Usually the points of attachment in females are structurally complementary to the shape of the male appendages so that pairing and subsequent mating between different species is avoided. Once attached, a male and female pair can fly around together—the male in front of the female [116]. Given that tandem arrangement, one might wonder how mating is possible.

A pair of Rambur's Forktail in a mating wheel. This species seems to be salt tolerant because it is most common at ponds near the coast such as the one by Route 9 near Woodland Beach shown above.

Mating in dragonflies and damselflies differs distinctly from that in other flying insects. Rather than direct transfer of sperm from the male to the female, there is an intermediate step where a sperm packet is transferred from the tip of the male's abdomen to a receptacle at the base of his abdomen near the thorax. From there, the sperm gets transferred to the female. Depending on the species, this may happen in a few seconds or last more than an hour. When complete, the pair may separate or the male may escort the female to a suitable oviposition site while remaining attached [86].

During mating, the female curls her abdomen under the male and connects with a complicated structure holding the sperm on the underside of the male's abdomen. This cyclic connection has been called the mating wheel, although some romantically inclined biologists have noted that the shape in side view looks more like a heart. Other biologists have speculated about how this unusual mating system might have evolved (20).

The figure shows a pair of Rambur's Forktail in a mating wheel. The male has green thorax and an all-blue eighth abdominal segment. The blue extends to the underside of the terminal ninth and tenth segments. The female has a dark abdomen and a greenish brown thorax.

Although the Rambur's Forktail occurs throughout the Delmarva Peninsula, it seems to be most common around freshwater ponds near salt marshes, implying that they are more salt tolerant than many other species. A reliable place to find them is a pond beside Route 9 near Woodland Beach, Delaware. The small pond on the east side of the road is easy to find because it is at a sharp bend in the road near a metal observation tower used for bird watching at Woodland Beach State Wildlife Area. Rambur's Forktails fly there from May through October.

125 MEASURING TIME
Ischnura verticalis (Eastern Forktail)

Imagine that you are a damselfly larva that has spent the winter under the ice. The entire winter your watery environment has not gone below freezing even though the air temperature above may have stayed below freezing for days at a time. Now it is spring. The water is warming up, but there is still the threat of frost in the thermally unbuffered world above. How do you know when it is safe to emerge? What cues do you have?

Not surprisingly, just as different kinds of plants leaf out or flower at characteristic times each year, dragonflies and damselflies emerge at different times as well; and they use the same cue—day length. While the air temperatures may vary greatly and unpredictably as they warm toward summer, the length of the day increases very predictably every year. Farmers use the calendar to determine when the frost-free season normally starts. Whatever date that is, it has a particular day length associated with it. Individual species may be less cold tolerant than others and have a later emergence date (day-length) that is optimal for them.

The details of how insects measure time in biochemical ways involves photosensitive molecules and is an area of active research on fruit flies. One assumes by analogy that the same process with slight variations operates in other insects. Thus, we have a general idea of how the larvae of the Eastern Forktail "know" when it is time to emerge. It is one of the first species to appear in early April and challenges the early appearance dates for the immigrating, much larger Green Darners [5]. Like the Green Darner, the Eastern Forktail has a long flight season and may persist until the first killing frost. However, the individuals found in the fall are not the same individuals that emerged in the spring. With a short life cycle and favorable conditions, they may squeeze in three generations a year.

Female Eastern Forktails normally mate just once and carry up to 2000 eggs. As with other forktail species, females come in different color forms [121], but the gray form is most common. Males are mostly black with bright green markings on the thorax and a bright blue tip to their abdomen with a black spot on the each side of the two blue segments. While they prefer ponds, they seem to turn up almost anywhere including the backwaters of streams and rivers. Eastern Forktails are common and widespread in the northeastern United States. However, they have not been found yet on the Virginia portion of Delmarva.

HAL WHITE

JIM WHITE

The male Eastern Forktail eating a small fly (top) has green markings on its thorax and a blue tip to its abdomen. Most female Eastern Forktails (bottom) become blue-gray as they age.

126 PURGATORY SWAMP
Nehalennia gracilis (Sphagnum Sprite)

Few things will hold my attention more than a United States Geological Survey topographic map. Long before GPS, geocaching, and GoogleEarth, I bushwhacked to destinations generated by long hours of map study. The forecast of sunny weather heightened the exercise of nightly map scanning as the weekend approached. For others, it is the anticipation of catching a trophy fish or a round of golf. For me, it is the hope of locating a pristine aquatic habitat swarming with dragonflies that one never sees at the local algae-choked pond.

As one learns to read topographic maps, the lay of the land becomes apparent. One can deduce that a stream must flow faster and be wadeable with a rocky bottom in some areas, and be deep and mucky in other places. The dragonflies on different parts of the stream will be different. One learns of places with character names given long ago like Purgatory Swamp or Hardscrabble Hill—nothing vapid like Hilltop Garden Heights, in a place with no hills in sight, or Wooddale Farms, where developers have bulldozed trees to plant houses.

In Massachusetts, I discovered that there were numerous Purgatory Swamps on my maps. I imagine Colonial settlers used that name to keep people out

Despite the iridescent, emerald-green top to its thorax and bright blue tip to its abdomen, the male Sphagnum Sprite rarely seen unless seekers are willing to explore bogs and fens.

because they might get stuck and never return. I confess that I have explored at least two and returned to tell my tale. I would return again to these wonderful sphagnum bogs that do in fact warrant careful steps. They rarely swarm with Odonates; but those that are there, are usually notable. One is the Sphagnum Sprite, a tiny damselfly that may be abundant yet goes unnoticed. Unfortunately, Delmarva's own Purgatory Swamp in Delaware near Cooch's Bridge south of Newark, is not a sphagnum bog now and thus does not have Sphagnum Sprites. Whether it did when the British Army bogged down in it on their way to take Philadelphia in 1777 is unknown.

The Sphagnum Sprite lives in sphagnum-choked bogs and fens [98]. That there are only a handful of such habitats known on Delmarva, and those often cover less than an acre, contributes to the rarity of Sphagnum Sprites. Nevertheless, at these sites they may be common in June and July. They fly and perch among grasses in wet sphagnum where they are well-camouflaged by their slender, iridescent green bodies that are less than an inch long. The tip of the abdomen is bright blue. Unlike other sprites, the blue area of the male Sphagnum Sprite lacks dark spots on the side. The female does have dark spots on the blue sides of the eighth and ninth segments, but the last, or terminal, segment is entirely blue.

127 SEEING THE UNEXPECTED

Nehalennia integricollis (Southern Sprite)

The Southern Sprite rivals the Citrine Forktail [120] as our smallest damselfly. This tiny damselfly, with its iridescent emerald green body and blue-tipped abdomen, flies low among the stems of emergent vegetation. Because it is inconspicuous and rarely flies unless disturbed, it is easy to overlook even when searching for it. The presence of one Cyrano Darner [14] at a pond attracts attention while hundreds, or even thousands, of Southern Sprites might be at the same location and go unnoticed.

I discovered my first Southern Sprite on the Delmarva Peninsula in 2002 after living here for more than three decades. Having found one, I looked harder and found a few more at a location that I had explored several times before without seeing one. Our minds are conditioned to see what we know and expect to see. We develop search images and look for the expected [111]. We return to the same places where we found something before rather than explore new places. I wonder how many times a Southern Sprite has seen me but I didn't see it because I didn't know what to look for or wasn't alert to the unexpected.

When groups of Odonate enthusiasts get together in the field, I am always amazed and interested in the species that others find that I did not see. The reason I missed them is often that I did not look in the right places at the right time. Dragonflies and damselflies do not distribute themselves evenly across a habitat. Rather, they prefer one microhabitat over others. Thus, one needs to know where to look. Furthermore, the activity patterns of a species may cause it to escape notice [50]. Some species are more active earlier in the day while others may be most active at dusk. They may forage in nearby fields before converging on breeding sites. There is so much knowledge of behavior, habitat, and geography that goes into finding rare or localized species. This is also why it is difficult to declare a species extinct. All we can say is that no one has seen it for a while.

I am still learning about the habits and habitat of the Southern Sprite. I have only a few data points to go on. My antennae go up at well-vegetated ponds or the backwaters of slow-flowing streams on the Coastal Plain in sandy environments. That the habitats may be a bit boggy means that one must be able to distinguish the Southern Sprite from the bog-loving Sphagnum Sprite [126]. Only the last abdominal segment of the Southern Sprite is entirely blue whereas the last two and most of the next segment are blue in the male Sphagnum Sprite. Look for Southern Sprites from June through mid-August.

The Southern Sprite is easy to miss because of its small size and its habit of perching in dense emergent vegetation in the shade. Consequently, its distribution on Delmarva is likely to be more extensive than indicated on the map.

128 RED ON GREEN
Telebasis byersi (Duckweed Firetail)

Adult dragonflies and damselflies are large-eyed, daytime predators. Their diet consists primarily of smaller flying insects. They feed near the top of the invertebrate food chain both as larvae and as adults and eat what they can catch. Even beautiful delicate damselflies are carnivores. Other insects, such as most butterflies and many beetles, feed on plants and often become associated with particular plants. For example, the Monarch butterfly is also called the Milkweed Butterfly because its larvae feed on milkweed. Similarly, we have cucumber and potato beetles, tomato hornworms, hickory horned devils, corn earworms, and fig wasps. Thus, to have a carnivorous damselfly, the Duckweed Firetail, named for its association with a plant should elicit curiosity.

Only a few dragonflies or damselflies have a plant as part of their name. Clearly, there must be other reasons than for food. The Spatterdock Darner (*Rhionaeshna mutata*), which has not yet been recorded on the Delmarva Peninsula, but could be found here, seems to prefer laying its eggs in the stems of spatterdock flowers. Consequently, any shallow pond with lots of spatterdock is potential habitat. In other cases, the dominant plant may simply be an indicator of a particular type of aquatic habitat where the species is often found as with the Sphagnum Sprite [126]. In the case of the Duckweed Firetail, there is more than a passing association with duckweed that makes the name quite appropriate.

A male Duckweed Firetail perches on duckweed in an abandoned sand pit in Kent County, Delaware. Despite the contrasting colors, this small red damselfly can be overlooked at a pond covered by a sea of duckweed.

The Duckweed Firetail, originally described from Florida (*94*), reaches its known northern limit of distribution on the East Coast in Kent County, Delaware. There it is found at a small, abandoned sand pit where the water becomes covered by a confluent mat of floating duckweed by mid-summer. Despite the male's brilliant red abdomen, it is sometimes hard to spot the small Duckweed Firetails perched directly on the duckweed in the sea of green. Females are brownish gray and better camouflaged. The larvae live in the duckweed mat, so the association with duckweed extends throughout the life cycle. Other than the Eastern Red Damsel [96], which lives in vegetation around spring seeps, a quite different habitat, there are no other bright red damselflies in our area. The Duckweed Firetail is also known from the Maryland and Virginia portions of Delmarva and flies in June and July.

129 METALLIC GEM AMONG EMERALDS

Somatochlora georgiana (Coppery Emerald)

In August 1971, Chris Leahy, now of the Massachusetts Audubon Society, brought me an unusual dragonfly he had found floating in a pond in Ipswich, Massachusetts. It was unlike any species I had ever seen before and it sent us scrambling for an identification. There were no species known from the northeastern United States like it. Our tentative identification of a Coppery Emerald seemed preposterous given that it was not known beyond five contiguous southeastern states, the closest of which was South Carolina nearly 1000 miles away! Yet that is what it was (*111*).

Subsequently, several populations of the Coppery Emerald have been found in eastern Massachusetts, southern New Hampshire, and Rhode Island. Isolated specimens also have been found in New Jersey and Virginia, so its distribution includes most of the east coast, but it is extremely rare. Thus, it was expected to be found on the Delmarva Peninsula sooner or later. Yet, as the years passed without discovery, its presence here seemed doubtful. As one observer put it, finding a Coppery Emerald on Delmarva was like seeking the Holy Grail.

The Holy Grail was found only weeks before this book manuscript went to the publisher, which explains why it appears here at the end rather than in its expected place. Dan Bogar visiting from Pennsylvania found a recently emerged brown dragonfly on the flood plain of the Marshyhope Creek in Caroline County, Maryland. He correctly identified it and deservedly gets the notoriety for his discovery.

The Coppery Emerald does not look like its emerald relatives. It is smaller than all of them, is copper brown rather than black or dark brown, and lacks the emerald green eyes one might expect. In fact, it does not look like an emerald at all. It is about the size of a baskettail, but more slender. Both of my previous encounters with the species occurred in New England in the late afternoon or near dusk as they flew with other dragonflies in feeding swarms. They flew high well out of net range most of the time, a behavior that may be the reason Coppery Emerald's escape detection.

As is often the case when someone makes a new discovery, others have an idea of where to look and find the Holy Grail elsewhere. In August 2010, Rick Cheicante discovered ovipositing Coppery Emeralds in the flood plain swamps of the Pocomoke watershed in two more counties of Eastern Maryland.

HAL WHITE

The recently emerged female Coppery Emerald above was found on June 18, 2010, by Dan Bogar at the Idylwild Wildlife Management Area in Caroline County, Maryland. It is the first record ever of this species on the Delmarva Peninsula.

APPENDIX I
Odonata species that might occur on the Delmarva Peninsula

This list contains species that probably are or were residents of the Delmarva Peninsula. In addition, species known to migrate long distances and may appear here sporadically are included. However, a number of species known from the mountains of Pennsylvania, New Jersey, and Maryland are not listed because suitable habitats for them are not known here.

Aeshna constricta	Marshy ponds, known from NJ, PA, & MD
Aeshna clepsydra	Known from southern NJ
Rhionaeshna mutata	Prefers fishless ponds with spatterdock, known from NJ, PA, & MD
Gomphus quadricolor	Known from Howard Co. Maryland
Gomphus vastus	Known from the Susquehanna River across from Cecil Co. MD
Ophiogomphus rupinsulensis	Known from Howard Co. Maryland
Macromia alleghaniensis	Coastal plain streams, known from southern NJ
Helocordulia uhleri	Occurs on clean flowing Piedmont Streams in NJ, PA, & MD
Dorocordulia lepida	Well known from southern New Jersey
Celithemis ornata	Southern, but reported from NJ
Ladona exusta	Common in NJ Pine Barrens
Libellula auripennis [59]	Known from PA, MD, and NJ
Sympetrum corruptum [77]	Western migrant, sporadic records along Atlantic coast to Nova Scotia
Sympetrum obtrusum	Known from southern NJ
Miathyria marcella	Tropical migrant along coast, known from VA
Macrodiplax balteata	Tropical migrant along coast, known from NC
Lestes unguiculatus	Well vegetated ponds, known from southern NJ
Enallagma pictum	Locally common in the NJ pine barrens, but not known to the south
Enallagma recurvatum	Early season, unpolluted ponds in NJ and Long Island sand barrens

APPENDIX II
Historic records for Odonata on the Delmarva Peninsula

The following species included in the text are known from isolated historic records

Tachopteryx thoreyi [1]	Hillside seeps; quite local. Known from Cecil Co., MD (1938)
Aeshna tuberculifera [2]	Marshy ponds; known from New Castle County, DE (1982)
Aeshna verticalis [4]	Marshy ponds, known from New Castle County, DE (1982)
Ophiogomphus incurvatus [23]	Piedmont streams like Big Elk, White Clay, and Brandywine creeks (1973)
Epitheca semiaquea [39]	Well known in NJ Pine Barrens and south of Delmarva; Delmarva records unconfirmed.
Leucorrhinia intacta [58]	New Castle Co. DE (1972)
Argia sedula [101]	Common in Potomac River basin, Brandywine Creek (1944)

APPENDIX III
Odonata habitats on the Delmarva Peninsula

Each year, the first dragonflies and damselflies begin to emerge as adults about mid April and the last few species disappear only after the first hard freeze in November. Some species occur early in the year, others late, and some fly throughout the summer. The best strategy for seeing the greatest variety of dragonflies and damselflies is to locate several different habitats and visit them through the flying season, preferably on warm, sunny days. The following are some habitat types to consider.

MILL PONDS—The Delmarva Peninsula has many old mill ponds that dot road maps and sometimes provide the centerpiece for a state park. Many of these ponds have public access points where one can see many of the common pond species or can launch a boat and explore various microhabitats for more elusive species.

TAILINGS PONDS—Several ponds and marshes exist along both sides of the Chesapeake & Delaware Canal between Route 896 in Delaware and Route 213 in Maryland. These wetlands often harbor distinctive populations of Odonata.

ABANDONED SAND BORROW PITS—Topographic maps and aerial photographs on GoogleEarth show many sand mining operations on the Delmarva Peninsula. Most of these are on private land. Abandoned pits often lack fish and have species found infrequently elsewhere. Perhaps the best accessible ponds of this type are found in the Idylwild Wildlife Management Area of Caroline County in Maryland.

DELMARVA BAYS—Depressions in the landscape that fill with water and normally dry up by July or August dot the landscape across the Coastal Plain. The more persistent of these support significant Odonata populations. The Whale Wallow at Lums Pond State Park and the Tybout Delmarva Bay in Blackbird State Forest in Delaware are two such places on public lands.

BOGS AND SEA-LEVEL FENS—In contrast to the Pine Barrens of New Jersey that have many bogs, the Delmarva Peninsula on the other side of the Delaware Bay has very few. These are small and inaccessible for the most part. They tend to occur where ground water seeps out of sandy soil just above the high tide and supports a mat of sphagnum moss and other bog plants. Several such bogs with characteristic bog dragonflies and damselflies occur in the Primehook National Wildlife Refuge.

LARGE RIVERS—The Susquehanna River forms the western border of Cecil County, Maryland. Access is limited and sometimes dangerous due to tidal changes, steep banks, and rapid flow below the Conowingo Dam. Several riverine species occur here and others may be found.

PIEDMONT STREAMS—A number of medium-sized streams have carved the valleys among the hills of northern Cecil and New Castle Counties. These include Brandywine Creek, White Clay Creek, Big Elk Creek, and Octoraro Creek. All of these flow through public lands and have access areas. The abundance of Odonata on these rocky-bottomed streams is considerably less than at ponds, but the species are different.

COASTAL PLAIN STREAMS—Streams on the Coastal Plain tend to flow slowly and have sandy, mud, or silt bottoms. The flood plains are often broad, wet, and difficult to navigate due to thick undergrowth. Thus, the best access points are at bridge crossings. The Choptank River at Christian Park near Goldsboro, Unicorn Creek at the Fish Hatchery near Millington, and various access areas along the Nanticoke and Pocomoke Rivers provide good places to observe Odonata.

TIDAL FRESH WATER STREAMS—Some of the larger streams on the Delmarva Peninsula like the Christina, the Choptank, the Nanticoke, and the Pocomoke are tidal in their lower reaches. These sections are virtually inaccessible without a boat. There are at least two clubtail species that have been found only in tidal fresh water.

SPRING SEEPS—The Piedmont region of northern Cecil and New Castle Counties have numerous spring seeps. Few Odonates are found there, but those that are tend to be found nowhere else.

SALT MARSHES—It has been said that 10% of the land area of the Delmarva Peninsula disappears twice a day at high tide. Extensive salt marshes exist along the Delaware and Chesapeake Bays. Several species abundant there are found nowhere else. Route 9 in Delaware crosses several salt marshes and tidal streams along the Delaware Bay and provides good access to these otherwise inaccessible habitats.

APPENDIX IV Delmarva Dragonfly Checklist by County

GENUS	SPECIES	New Castle	Kent, DE	Sussex	Cecil	Kent, MD	Queen Annes	Caroline
Tachopteryx	thoreyi				×			
Aeshna	tuberculifera	×						
Aeshna	umbrosa	×			×			×
Aeshna	verticalis	×						
Anax	junius	×	×	×	×	×		
Anax	longipes	×		×	×			×
Basiaeschna	janata	×	×	×	×	×		
Boyeria	vinosa	×	×	×	×	×	×	
Coryphaeschna	ingens							
Epiaeschna	heros	×	×	×	×			
Gomphaeschna	antilope			×				×
Gomphaeschna	furcillata	×				×		
Gynacantha	nervosa	×						
Nasiaeschna	pentacantha	×	×	×				
Arigomphus	villosipes	×		×	×		×	
Dromogomphus	spinosus	×	×	×	×	×	×	
Gomphus	apomius			×				
Gomphus	exilis	×	×	×	×	×		×
Gomphus	fraternus			×				
Gomphus	lividus	×	×	×				
Gomphus	rogersi	×					×	
Hagenius	brevistylus	×	×	×	×			
Progomphus	obscurus	×	×				×	
Stylogomphus	albistylus	×			×			
Stylurus	laurae		×					
Stylurus	plagiatus	×		×				×
Stylurus	spiniceps	×			×			
Cordulegaster	bilineata	×						
Cordulegaster	diastatops	×		×			×	
Cordulegaster	erronea	×			×			×
Cordulegaster	maculata	×	×	×	×			
Cordulegaster	obliqua							×
Didymops	transversa	×	×	×		×	×	
Macromia	georgina	×					×	×
Macromia	illinoiensis	×			×			
Macromia	taeniolata	×	×	×				×
Epitheca	costalis	×		×			×	
Epitheca	cynosura	×	×	×	×	×	×	×
Epitheca	princeps	×		×	×	×	×	×
Epitheca	semiaquea			×				
Epitheca	spinosa			×				×
Helocordulia	selysii		×					×
Neurocordulia	obsoleta			×	×		×	×
Neurocordulia	yamaskanensis				×			

266

Talbot	Dorchester	Wicomico	Worchester	Somerset	Accomack	Northampton	COMMON NAME	DATES
							Grey Petaltail	7/7
							Black-tipped Darner	10/3
		×					Shadow Darner	8/26-11/15
							Green-striped Darner	9/13-10/3
×	×	×	×	×	×	×	Common Green Darner	4/5-11/9
×	×	×	×	×		×	Comet Darner	5/16-8/7
	×	×	×	×			Springtime Darner	4/10-6/9
	×	×	×		×		Fawn Darner	7/4-10/31
						×	Regal Darner	7/26
×	×	×	×	×	×	×	Swamp Darner	4/27-9/8
		×	×				Taper-tailed Darner	5/20-7/1
	×	×	×			×	Harlequin Darner	4/11-6/7
							Twilight Darner	9/27
		×	×				Cyrano Darner	5/27-8/31
	×	×	×	×			Unicorn Clubtail	5/22-7/23
	×						Black-shouldered Spinyleg	5/28-9/6
							Banner Clubtail	5/4-6/2
	×	×	×	×			Lancet Clubtail	4/12-7/12
							Midland Clubtail	5/13-6/7
	×	×	×				Ashy Clubtail	4/11-7/4
							Sable Clubtail	5/17-6/27
		×	×				Dragonhunter	5/31-9/11
	×	×	×				Common Sanddragon	5/27-7/26
							Least Clubtail	6/9-6/27
							Laura's Clubtail	6/17
×			×				Russet-tipped Clubtail	8/18-10/13
						×	Arrow Clubtail	6/20-9/9
							Brown Spiketail	6/6-7/5
					×		Delta-spotted Spiketail	6/1-6/2
							Tiger Spiketail	6/26-8/31
	×	×					Twin-spotted Spiketail	4/10-5/20
							Arrowhead Spiketail	5/29
		×	×				Stream Cruiser	4/11-6/7
	×	×	×				Georgia River Cruiser	6/7-9/2
							Illinois River Cruiser	6/9-9/4
		×	×		×		Royal River Cruiser	6/25-9/6
							Striped-wing Baskettail	6/2-6/19
	×	×	×	×			Common Baskettail	4/11-7/7
	×	×			×		Prince Baskettail	5/27-10/16
							Mantled Baskettail	4/27-5/10
		×					Robust Baskettail	4/11-5/9
							Sely's Sundragon	4/11-5/4
							Umber Shadowdragon	5/27-7/2
							Stygian Shadowdragon	6/2-7/2

GENUS	SPECIES	New Castle	Kent, DE	Sussex	Cecil	Kent, MD	Queen Annes	Caroline	
Somatochlora	*filosa*	×	×	×		×		×	
Somatochlora	*georgiana*							×	
Somatochlora	*linearis*	×	×	×		×		×	
Somatochlora	*provocans*			×				×	
Somatochlora	*tenebrosa*	×	×	×			×	×	
Brachymesia	*gravida*			×		×	×		
Celithemis	*elisa*	×	×	×	×	×	×	×	
Celithemis	*eponina*	×	×	×	×	×	×	×	
Celithemis	*fasciata*			×				×	
Celithemis	*martha*			×					
Celithemis	*verna*	×		×	×			×	
Erythemis	*simplicicollis*	×	×	×	×	×	×	×	
Erythrodiplax	*berenice*	×	×	×		×	×	×	
Erythrodiplax	*minuscula*							×	
Ladona	*deplanata*	×	×	×	×	×		×	
Leucorrhinia	*intacta*	×							
Libellula	*axilena*	×		×	×			×	
Libellula	*cyanea*	×	×	×	×	×	×	×	
Libellula	*flavida*	×			×			×	
Libellula	*incesta*	×	×	×	×	×	×	×	
Libellula	*luctuosa*	×	×	×	×	×	×	×	
Libellula	*needhami*	×	×	×	×	×		×	
Libellula	*pulchella*	×	×	×	×	×	×	×	
Libellula	*semifasciata*	×	×	×	×			×	
Libellula	*vibrans*	×	×	×	×	×	×	×	
Nannothemis	*bella*	×	×	×				×	
Orthemis	*ferruginea*								
Pachydiplax	*longipennis*	×	×	×	×	×		×	
Pantala	*flavescens*	×	×	×	×	×	×	×	
Pantala	*hymenaea*	×	×	×	×	×	×	×	
Perithemis	*tenera*	×	×	×	×			×	
Plathemis	*lydia*	×	×	×	×	×	×	×	
Sympetrum	*ambiguum*	×		×	×			×	
Sympetrum	*janeae*	×	×		×				
Sympetrum	*rubicundulum*	×		×	×			×	
Sympetrum	*semicinctum*	×			×				
Sympetrum	*vicinum*	×	×	×	×	×	×	×	
Tramea	*calverti*		×						
Tramea	*carolina*	×	×	×	×	×	×	×	
Tramea	*lacerata*	×	×	×	×	×	×	×	
Tramea	*onusta*								
TOTALS	**85**	**64**	**44**	**61**	**48**	**35**	**43**	**59**	

10/10/2010

Talbot	Dorchester	Wicomico	Worchester	Somerset	Accomack	Northampton	COMMON NAME	DATES
	×	×	×				Fine-lined Emerald	7/27-9/8
		×	×				Coppery Emerald	6/12-8/31
		×	×				Mocha Emerald	6/10-9/20
							Treetop Emerald	6/18-8/1
							Clamp-tipped Emerald	6/10-9/3
	×		×	×	×	×	Four-spotted Pennant	6/19-9/9
×	×	×	×	×			Calico Pennant	5/27-10/10
×	×	×	×	×	×	×	Halloween Pennant	6/9-10/3
		×					Banded Pennant	6/12-9/4
		×					Martha's Pennant	7/20
		×					Double-ringed Pennant	6/12-7/24
×	×	×	×	×	×	×	Eastern Pondhawk	5/7-10/15
×	×	×	×	×	×	×	Seaside Dragonet	5/22-10/9
							Little Blue Dragonlet	7/24
	×	×	×	×		×	Blue Corporal	4/11-6/12
							Dot-tailed Whiteface	7/9
	×	×	×		×		Bar-winged Skimmer	5/30-9/1
×	×	×	×				Spangled Skimmer	5/7-8/21
							Yellow-sided Skimmer	5/21-8/1
×	×	×	×	×			Slaty Skimmer	5/20-10/15
×	×	×	×	×		×	Widow Skimmer	5/28-10/9
×	×	×	×	×	×		Needham's Skimmer	5/21-9/20
×	×	×	×	×	×	×	Twelve-spotted Skimmer	5/13-10/13
×	×	×	×	×	×		Painted Skimmer	4/16-8/8
	×	×	×		×	×	Great Blue Skimmer	5/22-10/1
							Elfin Skimmer	5/21-8/4
						×	Roseate Skimmer	10/15
×	×	×	×	×	×	×	Blue Dasher	5/9-10/22
×	×	×	×	×		×	Wandering Glider	5/20-10/22
×	×	×	×		×	×	Spot-winged Glider	5/11-10/7
×	×	×	×	×		×	Eastern Amberwing	4/12-10/3
×	×	×	×	×	×		Common Whitetail	4/12-10/15
			×		×	×	Blue-faced Meadowhawk	6/5-11/9
							Cherry-faced Meadowhawk	6/21-9/26
							Ruby Meadowhawk	6/14-10/3
							Band-winged Meadowhawk	6/8-9/12
×			×	×			Autumn Meadowhawk	6/24-12/01
		×					Stripped Saddlebags	8/11-8/22
×	×	×	×	×	×		Carolina Saddlebags	5/16-10/9
×	×	×	×	×	×		Black Saddlebags	5/20-10/9
		×					Red Saddlebags	
22	37	40	53	27	22	19		

Maryland County records compiled by Richard Orr. Virginia County records compiled by Steve Roble

APPENDIX V — Delmarva Damselfly Checklist by County

GENUS	SPECIES	New Castle	Kent, DE	Sussex	Cecil	Kent, MD	Queen Annes	Caroline
Calopteryx	dimidiata		×	×			×	×
Calopteryx	maculata	×	×	×	×	×	×	×
Hetaerina	americana	×		×	×			
Archilestes	grandis	×						
Lestes	australis	×		×	×	×	×	×
Lestes	congener	×				×		×
Lestes	eurinus	×						
Lestes	forcipatus	×	×	×			×	×
Lestes	inaequalis	×	×	×	×			×
Lestes	rectangularis	×	×	×	×			×
Lestes	vigilax	×	×	×				×
Amphiagrion	saucium	×		×			×	
Argia	apicalis	×	×	×	×	×	×	×
Argia	bipunctulata		×	×				
Argia	violacea	×	×	×	×	×		×
Argia	moesta	×			×			
Argia	tibialis	×	×	×		×		
Argia	translata	×			×			
Chromagrion	conditum	×		×	×			×
Enallagma	aspersum	×	×	×			×	×
Enallagma	basidens	×	×	×			×	×
Enallagma	civile	×		×	×	×		×
Enallagma	daeckii		×	×				×
Enallagma	divagans	×		×			×	×
Enallagma	doubledayi							
Enallagma	dubium			×				×
Enallagma	durum	×			×	×		
Enallagma	exsulans	×	×	×	×		×	×
Enallagma	geminatum	×		×	×	×		×
Enallagma	pallidum			×				×
Enallagma	signatum	×	×	×	×	×	×	×
Enallagma	traviatum	×	×	×			×	×
Enallagma	vesperum	×		×			×	×
Enallagma	weewa			×			×	×
Ischnura	hastata	×	×	×	×	×	×	×
Ischnura	kellicotti		×	×				×
Ischnura	posita	×	×	×	×	×	×	×
Ischnura	prognata							×
Ischnura	ramburii	×	×	×			×	×
Ischnura	verticalis	×	×	×	×	×	×	×
Nehalennia	gracilis			×				×
Nehalennia	integricollis		×					×
Telebasis	byersi		×					×
TOTALS	**43**	**31**	**24**	**34**	**19**	**17**	**22**	**35**

Talbot	Dorchester	Wicomico	Worchester	Somerset	Accomack	Northampton	COMMON NAME	DATES
		×					Sparkling Jewelwing	5/20-7/27
	×	×	×	×	×		Ebony Jewelwing	5/3-9/8
							American Rubyspot	8/7-9/28
							Great Spreadwing	7/9-10/22
	×	×	×		×		Southern Spreadwing	5/2-11/5
							Spotted Spreadwing	7/2-11/8
							Amber-winged Spreadwing	5/31-6/22
	×						Sweetflag Spreadwing	5/30-10/15
			×		×		Elegant Spreadwing	7/6-8/26
	×		×		×	×	Slender Spreadwing	5/12-10/15
		×	×				Swamp Spreadwing	6/8-9/6
	×	×			×		Eastern Red Damsel	4/30-7/5
×			×				Blue-fronted Dancer	6/4-10/21
					×		Seepage Dancer	6/25-8/4
	×	×	×		×		Violet Dancer	5/27-10/2
							Powdered Dancer	5/22-9/19
		×	×				Blue-tipped Dancer	5/27-9/5
							Dusky Dancer	7/15-9/4
		×	×				Aurora Damsel	5/19-6/27
×	×	×	×		×		Azure Bluet	4/7-9/1
							Double-striped Bluet	5/12-9/13
×		×	×	×	×		Familiar Bluet	4/30-11/9
		×	×		×		Attenuated Bluet	5/27-7/24
	×	×	×	×			Turquoise Bluet	5/13-7/14
		×					Atlantic Bluet	7/18-9/9
			×				Burgundy Bluet	6/15-8/31
	×						Big Bluet	6/9-10/21
			×				Stream Bluet	6/9-10/1
	×	×	×		×		Skimming Bluet	5/1-9/25
		×					Pale Bluet	7/10-7/24
	×	×	×	×			Orange Bluet	4/29-9/20
		×					Slender Bluet	5/30-8/18
		×					Vesper Bluet	5/9-8/4
		×	×	×			Blackwater Bluet	5/19-8/31
×	×	×	×	×	×	×	Citrine Forktail	4/17-11/29
		×		×	×		Lilypad Forktail	5/4-9/2
×	×	×	×	×	×	×	Fragile Forktail	4/3-10/15
		×	×		×		Furtive Forktail	5/19-7/15
×	×	×	×	×	×	×	Rambur's Forktail	5/18-11/9
×		×	×	×			Eastern Forktail	4/12-10/16
		×	×				Sphagnum Sprite	5/26-8/4
		×					Southern Spite	5/27-7/29
		×	×		×		Duckweed Firetail	7/6-7/25
7	14	26	26	10	16	5		

REFERENCES

REF. NO.   CITATION   VIGNETTE NUMBER

1. Bagg, Aaron M. 1958. Fall Emigration of the Dragon-fly *Anax junius*. *Maine Field Naturalist* 14: 2-13.　　5

2. Barber, Robert D. 1994. *Damselflies and Dragonflies of Cumberland County, New Jersey*. Cape May, NJ: Cape May Bird Observatory.　　**39, 62, 110**

3. Barlow, Allen E., David M. Golden, and Jim Bangma. 2009. *Field Guide to the Dragonflies and Damselflies of New Jersey*. New Jersey Department of Environmental Protection, Division of Fish and Wildlife.　　**Preface, 39**

4. Beaton, Giff. 2007. *Dragonflies and Damselflies of Georgia and the Southeast*. Athens, GA. University of Georgia Press.　　**Preface**

5. Beatty, George H., III. 1946. Dragonflies (Odonata) Collected in Pennsylvania and New Jersey in 1945. *Entomological News* 52: 1-10, 50-56, 76-81, 104-111.　　**2**

6. Beatty, George H. III, and Alice F. Beatty. 1963. Efficiency in Caring for Large Odonata Collections. *Proc. N. Cent. Br.-Ent. Soc. Am.* 18: 149-153.　　**14**

7. Beatty, George H. III, Alice F. Beatty, and Clark N. Shiffer. 1970. A Survey of the Odonata of Eastern Pennsylvania. *Pennsylvania Academy of Sciences Proceedings* 44: 141-152.　　**11, 41**

8. Beckemeyer, Roy. 2000. The Permian Insect Fauna of Elmo, Kansas. *The Kansas School Naturalist* 46(1).　　**20**

9. Bick, George H. and Juanda C. Bick. 1965. Demography and Behavior of the Damselfly, *Argia apicalis* (Say), (Odonata: Coenagrionidae). *Ecology* 46: 461-472.　　**97**

10. Bick, George H. and Juanda C. Bick. 1965. Color Variation and Significance of Color in Reproduction in the Dragonfly, *Argia apicalis* (Say) (Zygoptera: Coenagrionidae) *Canadian Entomologist* 97: 32-41.　　**97**

11. Blust, Michael H. 1980. The life history and production ecology of *Stylogomphus albistylus* (Hagen): (Odonata: Gomphidae). MS diss., University of Delaware.　　**25**

12. Borror, Donald J. 1963. Common Names for Odonata, *Proceedings of the North Central Branch, Entomological Society of America* 18: 104-107.　　**50**

13. Bree, David. 2001. Mantids eating dragonflies. *Argia* 13(1): 27-28.　　**71**

14. Brown, Jonathan M., Mark A. McPeek and Michael L. May. 2000. A phylogenetic perspective on habitat shifts and diversity in the North American *Enallagma* damselflies. *Systematic Biology* 49: 697-712.　　**114**

15. Calvert, Philip P. 1893. Dr. H. A. Hagen [obituary, with portrait]. *Entomological News* 4: 313-317.　　**107**

16. Calvert, Philip P. 1893. Catalog of the Odonata (Dragonflies) of the Vicinity of Philadelphia, with an Introduction to this Group of Insects. *Transactions of the American Entomological Society* 20: 152-272.　　**108**

17. Calvert, Philip P. 1903. Additions to the Odonata of New Jersey, with descriptions of two new species. *Entomological News* 14: 33-41.　　**46, 108**

18. Calvert, Philip P. 1926. Relations of an autumnal dragonfly (Odonata) to temperature. *Ecology* 7: 185-190.　　**80**

19. Carle, Frank Louis. 1982. *Ophiogomphus incurvatus*: A new name for *Ophiogomphus carolinus* Hagen (Odonata: Gomphidae). *Annals of the Entomological Society of America* 75: 335-339.   23

20. Carle, Frank Louis. 1982. Evolution of the odonate copulatory process. *Odonatologica* 11: 271-286.   124

21. Carle, Frank Louis. 1983. A new *Zoraena* (Odonata: Cordulegasteridae) from eastern North America, with a key to the adult Cordulegasteridae of America. *Annals of the Entomological Society of America* 76: 61-68.   17, 29

22. Carle, Frank Louis. 1993. *Sympetrum janeae* spec. nov. from eastern North America, with key to nearctic *Sympetrum* (Anisoptera: Libellulidae). *Odonatologica* 22: 1-16.   17, 78

23. Carson, Rachel. 1962. *Silent Spring*, New York.Houghton Mifflin.   51

24. Corbet, Philip S. 2000. The first recorded arrival of *Anax junius* Drury (Anisoptera: Aeshnidae) in Europe: A scientist's perspective. *International Journal of Odonatology* 3: 153-162.   6

25. Corbet, Philip S. and R. Hoess. 1998. Sex ratio of Odonata at emergence. *International Journal of Odonatology* 1: 99-118.   2

26. Donnelly, Thomas W. 1966. A new *Gomphine* dragonfly from Eastern Texas (Odonata: Gomphidae). *Proceedings of the Entomological Society of Washington* 68(2): 102-105.   17

27. Donnelly, Thomas W. 1973. The status of *Enallagma traviatum* and westfalli (Odonata: Coenagrionidae). *Proceedings of the Entomological Society of Washington* 75: 297-302.117

28. Donnelly, Nick [Thomas W.]. 1995. Do Dragonflies make Sound? And What on Earth For? *Argia* 7(1): 23-25.   43

29. Donnelly, Thomas W. 1998. History of American Odonata Study: E. B. Williamson. *Argia* 10(3): 10-13.   52

30. Donnelly, Thomas W. 2003. *Lestes disjunctus, forcipatus*, and *australis*: a confusing complex of North American damselflies. *Argia* 15(3): 10-11.   89

31. Donnelly, Thomas W. and Kenneth J. Tennessen. 1994. *Macromia illinoiensis* and *georgina*: A Study of their Variation and Apparent Subspecific Relationship (Odonata: Corduliidae) *Bulletin of American Odonatology* 2(3), 27-61.   34

32. Dunkle, Sidney W. 1977. Larvae of the genus *Gomphaeschna* (Odonata, Aeschnidae). *The Florida Entomologist* 60: 223-225.   12

33. Dunkle, Sidney W. 1981. The Ecology and Behavior of *Tachopteryx thoreyi* (Hagen) (Anisoptera: Petaluridae). *Odonatologica* 10: 189-199.   1

34. Dunkle, Sidney W. 1989. *Dragonflies of the Florida Peninsula, Burmuda and the Bahamas*. Gainesville, FL: Scientific Publishers.   13

35. Dunkle, Sidney W. 1991. Head damage from mating attempts in dragonflies (Odonata: Anisoptera). *Entomological News*102: 37-41.   22

36. Dunkle, Sidney W. 2000. *Dragonflies Through Binoculars: A Field Guide to the Dragonflies of North America*. New York: Oxford University Press.   78

37. Eisner, Thomas. 2004. *For Love of Insects*. Cambridge, MA: Belknap (Harvard University Press).   85

38. Feynman, Richard P. and Robert M. Leighton. 1988. *What Do You Care What Other People Think? Further Adventures of a Curious Character*. New York: W. W. Norton & Co. **Preface**

39. Fincke, Ola M., Reinhard Jödicke, Dennis R. Paulson, and Thomas D. Schultz. 2005. The evolution and frequency of female color morphs in Holarctic Odonata: Why are male-like females typically the minority? *International Journal of Odonatology* 8: 183-212. **2, 74**

40. Fisher, E. G. 1940. A list of Maryland Odonata. *Entomological News* 51: 37-42, 67-72. **1, 30**

41. Gibbs, Robert H., Jr. and Sarah P. Gibbs. 1954. The Odonata of Cape Cod, Massachusetts, *New York Entomological Society Proceedings* 62:167-184. **59**

42. Glotzhober, Robert C. and David McShaffrey, D. ed. 2002. The Dragonflies and Damselflies of Ohio. Columbus, OH: Ohio Biological Survey. **54**

43. Gloyd, Leonora K. 1968. The union of *Argia fumipennis* (Burmeister, 1839) with *Argia violacea* (Hagen, 1861), and the recognition of three subspecies (Odonata). *Occasional Papers of the Museum of Zoology, University of Mich*igan No 658: 1-6. **99**

44. Heckscher, Christopher M. and Harold B. White, III. 2005. First Atlantic Coastal Plain Occurence of *Gomphus fraternus* Say (Odonata; Gomphidae). *Entomological News*119: 167-168. **19**

45. Hegner, Robert. 1935. Francis Metcalf Root. *J. Parasitol.* 21: 66-69. **111, 115**

46. Jacobs, Merle E. 1955. Studies on territorialism and sexual selection in Dragonflies. *Ecology* 36: 566-586. **75**

47. Johnson, Cliffford. 1973. Variability and Taxonomy of *Calopteryx dimidiata* (Zygoptera: Calopterigidae). *The Florida Entomologist* 56: 207-222. **85**

48. Kaushal, Sujay S., Peter M. Groffman, Gene E. Likens, Kenneth T. Belt, William B. Stack, Victoria K. Kelly, Lawrence E. Band, and Gary T. Fisher. 2005. Increased salinization of fresh water in the northeastern United States. *Proceedings of the National Academy of Science U.S.A.* 102: 13517-13520. **112**

49. Lam, Edward. 2004. *Damselflies of the Northeast*. Forest Hills, NY: Biodiversity Books. **Preface**

50. May, Michael L. 1998. Body temperature regulation in a late season dragonfly, *Sympetrum vicinum* (Odonata: Libellulidae). *International Journal of Odonatology* 1: 1-13. **80**

51. McDermott, Frank A. 1963. Frank Morton Jones (1869-1962). *Entomological News* 74(2): 28-36. **41**

52. Montgomery, B. Ellwod. 1972. Why snakefeeders? Why dragonfly? Some random observations on etymological entomology. *Proceedings of the Indiana Academy of Science* 82: 235-241. **8**

53. Morrison, Fredriick, D. McLain, and Laurie Sanders. 2004. Dragonfly abundance and emergence behavior before and after bank stabilization on the Connecticut River in Gill, Massachusetts. Paper presented at the New England Odonata Conference, April 17, 2004, in Athol, MA. **18**

54. Murray, Molly. 1993. A fine fen-tastic find in Sussex. *Wilmington News Journal*, July 29, 1993, p B1. **98**

55. Nature Conservancy of Delaware. 2005. Blackbird-Millington Corridor Report. **21**

56. Needham, James G. and Hortense B. Heywood. 1929. *A Handbook of the Dragonflies of North America*. Springfield, IL: Charles C. Thomas. **50, 65**

57. Needham, James G., Minter J. Westfall, Jr., and Michael L. May. 2000. *Dragonflies of North America*, revised edition. Gainesville, FL: Scientific Publishers. **65, 111**

58. Nikula, Blair, Jennifer L. Loose, and Matthew R. Burn. 2003. *A Field Guide to the Dragonflies and Damselflies of Massachusetts*. Westborough, MA: Massachusetts Division of Fisheries and Wildlife, Natural Heritage and Endangered Species Program.   **Preface**

59. Norberg, R. Åke. 1972. The Pterostigma of Insect Wings: an Inertial Regulator of Wing Pitch. *Journal of Comparative Physiology* 81: 9—22.   **61**

60. Nye, David. 2002. Obituary in *Wilmington News Journal,* March 14, 2002.   **12**

61. Olberg, R., R. Seama, M. Coats, and A. Henry. 2007. Eye movements and target fixation during dragonfly prey-interception flights. *Journal of Comparative Physiology A* 19: 685-693.   **66**

62. Paulson, Dennis R. 1996. Sexism and Odonata Conservation, *Argia* 8(2): 31-34.   **2**

63. Paulson, Dennis R. 1999. Photo files for Odonate Records. *Argia* 11(3): 19-20.   **32**

64. Paulson, Dennis R. and Sidney W. Dunkle. 1999. A Checklist of North American Odonata Including English Name, Etymology, Type Locality, and Distribution. *Occasional Paper 56 of the Slater Museum of Natural History.* Tacoma, WA.   **50**

65. Pilgrim, E. M., S. A. Roush, and D. E. Krane. 2002. Combining DNA sequences and morphology in systematics: testing the validity of the dragonfly species *Cordulegaster bilineata. Heredity* 89. 184-190.   **29**

66. Pringle, Lawrence. (2001) *A Dragon in the Sky: The Story of a Green Darner Dragonfly*. New York: Orchard Books.   **5**

67. Cordero Rivera, A., M. O. Lorenzo Carballa, C. Utzeri, and V. Vieira. 2005. Parthenogenic *Ischnura hastata* (Say), Widespread in the Azores (Zygoptera: Coenagrionidae). *Odonatologica* 34: 1-9.   **120**

68. Root, Francis M. 1923. Notes on Zygoptera (Odonata) from Maryland, with a Description of *Enallagma pallidum*, n. sp. *Entomological News* 34: 200-204.   **115**

69. Root, Francis M. 1924. Notes on Dragonflies (Odonata) from Lee County, Georgia, with description of *Enallagma dubium*, New Species. *Entomological News* 35: 317-324. **109, 115**

70. Roache, Larry O., Judy M. Semroc, and Linda K. Gilbert. 2008. *Dragonflies and Damselflies of Northeast Ohio*, Second Edition, Cleveland, OH: Cleveland Museum of Natural History   **Preface**

71. Sawchyn, W. W. and C. Gillott. 1974. The Life History of *Lestes congener* (Odonata: Zygoptera) on the Canadian Prairies. *Canadian Entomologist* 106: 367-376.   **90**

72. Schaefer, Paul W., Susan E. Barth, and Harold B. White, III. 1996. Incidental Capture of Male *Epiaeschna heros* (Odonata: Aeshnidae) in Traps Designed for Arboreal *Calosoma sycophanta* (Coleoptera: Carabiidae). *Entomological News* 107: 261—266.   **10**

73. Schaefer, Paul W., Susan E. Barth, and Harold B. White, III. 1996 Predation on sweetpotato whitefly, *Bemisis tabaci* (Gennadius) (Homoptera: Aleyrodidae) by *Enallagma civile* (Hagen) (Odonata: Zygoptera: Coenagrionidae). *Entomological News* 107: 275—276.   **107**

74. Sherrat. Thomas N. and Christopher D. Beatty. 2005. Evolutionary Biology: Island of the clones. *Nature* 435: 1039-1040.   **118**

75. Shiffer, Clark N. 1968. Homeochromatic females in the dragonfly *Perithemis tenera. Proceedings of the Pennsylvania Academy of Sciences* 42:138-141.   **74**

76. Shiffer, Clark N. 1994. Old records of *Tramea calverti* from Maryland, *Argia* 6(3): 4.   **81**

77. Smith, Ian M., Bruce P. Smith, and David R. Cook. 2001. Water mites (Hydrachnida) and other arachnids. In *Ecology and Classification of North American Freshwater Invertebrates*, 2nd edition. ed. James H. Thorp and Alan P. Covich, 551-659. New York: Academic Press.   36

78. Soltesz, Kenneth. 1991. *A Survey of the Dragonflies and Damselflies of Cape May County, New Jersey*. Cape May, NJ: New Jersey Audubon Society.   110

79. Soltesz, Kenneth 1992. An Invasion of *Tramea calverti* on the Northeast Coast. *Argia* 4(3): 9-10.   81

80. Soltesz, Kenneth, Robert Barber, and Virginia Carpenter. 1995. A spring dragonfly migration in the Northeast. *Argia* 7(3): 10-14.   60

81. Stevenson, Dirk J., Steven M. Roble, and Christopher S. Hobson. 1995. New records of the damselfly *Ischnura prognata* in Virginia. *Banisteria* 6: 26-27.   123

82. Stroud Water Research Center stream surveys. Unpublished.   28

83. Stroud, Patricia Tyson. 1992. *Thomas Say-New World Naturalist*, Philadelphia, PA: University of Pennsylvania Press.   91

84. Tillyard, R. J. 1917. *The Biology of Dragonflies*. Cambridge, U.K.: Cambridge. University Press.   35, 105

85. U. S. Fish and Wildlife Service. 1999. *Hine's Emerald Dragonfly (Somatochlora hineana) Draft Recovery Plan*. Fort Snelling, MN.   65

86. Waage, Jonathan K. 1979. Dual function of the damselfly penis, sperm removal and transfer. *Science* 203: 916-918.   86

87. Walker, Edmund M. 1952. The *Lestes disjunctus* and *forcipatus* complex. *Transactions of the American Entomological Society* 78: 234-237.   89

88. Walker, Edmund M. 1953. *The Odonata of Canada and Alaska*, Vol 1. Toronto: University of Toronto Press.   89

89. Walker, Edmund M. 1958, *The Odonata of Canada and Alaska*, Vol 2. Toronto: University of Toronto Press.   89

90. Walker, Edmund M. 1966. On the generic status of *Tetragoneuria* and *Epicordulia* (Odonata: Corduliidae). *Canadian Entomologist* 98: 897-902.   40

91. Walker, Edmund M. and Philip S. Corbet 1978. *The Odonata of Canada and Alaska*, Vol 3. Toronto: University of Toronto Press.   89

92. Walter, Robert C. and Dorothy J. Merritts. 2008. Natural Streams and the Legacy of Water-Powered Mills. *Science* 319: 299-304.   100

93. Westfall, Minter J., Jr. 1943 The synonymy of *Libellula auripennis* Burmeister and *Libellula jesseana* Williamson, and a description of a new species, *Libellula needhami* (Odonata). *Transactions of the American Entomological Society* 69: 17-31.   65

94. Westfall, Minter J., Jr. 1957. A New Species of *Telebasis* from Florida (Odonata: Zygoptera). *Florida Entomologist* 40: 19-27.   128

95. Westfall, Minter J., Jr. and Michael L. May. 1996. *Damselflies of North America*. Gainesville, FL: Scientific Publishers.   111

96. Whigham, Dennis F. 1999. Ecological issues related to wetlands preservation, restoration, creation, and assessment. *Science of the Total Environment* 240: 312-40.   96

97. White, James F., Jr. and Amy W. White. 2007. *Amphibians and Reptiles of Delmarva,* Second Edition. Centreville, MD: Cornell Maritime Press, Tidewater. **Preface**

98. White, Harold B., III, and Rudolf A. Raff. 1970. Early Spring Emergence of *Anax junius* (Odonata: Aeshnidae) in Central Pennsylvania. *Canadian Entomologist* 102: 498-499.   5

99. White, Harold B., III. 1979. Notable instances of avoidance behavior in Odonata, *Notulae Odonatologica* 1: 67-69.   79

100. White, Harold B., III. 1984. Philip Powell Calvert: Student, Teacher, and Odonatologist. *Entomological News* 95: 155-162.   108

101. Wiggins, Glenn B. ed. 1966. *Centennial of Entomology in Canada 1863-1963: A Tribute to Edmund M. Walker.* Toronto: University of Toronto Press.   89

102. Wikelski, Martin, David Moskowitz, James S. Adelman, Jim Cochran, David S. Wilcove, and Michael L. May. 2006. Simple rules guide dragonfly migration. *Biology Letters* 2: 325-329.   5

103. Williamson, E. B. 1905. Oviposition in *Tetragoneuria* (Odonata). *Entomological News* 16: 255-7.   37

104. Williamson, E. B. 1922. Notes on *Celithemis* with descriptions of two new species (Odonata). *Occasional Papers of the Museum of Zoology, University of Michigan* 108: 1-22.   52

105. Williamson, E. B. 1931. Common Names for Dragonflies (Odonata). *Entomological News* 42: 46-50.   **50, 52**

106. Williamson, E. B. 1931. *Archilestes grandis* (Ramb.) in Ohio (Odonata:Agrionidae). *Entomological News* 42: 63-64.   88

107. Williamson, E. B. 1932. Two New Species of *Stylurus* (Odonata-Gomphinae). *Occasional Papers of the Museum of Zoology, University of Michigan* 247: 1-18.   26

108. Woodbury, Elton. N. 1994. *Butterflies of Delmarva.* Centreville, MD: Cornell Maritime Press, Tidewater.   **Preface**

109. Worthen, Wade B., Susan Gregory, Jason Felton, and Melisssa J. Hutton. 2004. Larval habitat associations of *Progomphus obscurus* at two spatial scales (Odonata: Gomphidae). *International Journal of Odonatology* 7: 97-109.   24

110. van Brink, Jan M. and Bastian Kiauta. 1977. To Mrs. Leonora K. Gloyd on her 75th Birthday. *Odonatologica* 6(3). 143-149.   101

111. White, Harold B., III, Paul Miliotis, and Christopher W. Leahy. 1974. Additions to the Odonata of Massachusetts. *Entomological News* 85: 208-210.   129

112. White, Harold B., III and John V. Calhoun. 2009. Miss Mattie Wadsworth (1862-1943), Early Woman Author in *Entomological News. Transactions of the American Entomological Society* 13: 413-429.   52

113. Blythe, R. H. 1950. *Haiku,* Hokuseido Press.   **53, 78**

INDEX

About the Delaware Nature Society

A nonprofit membership organization founded in 1964, the Delaware Nature Society fosters understanding, appreciation and enjoyment of the natural world through education; preserves ecologically significant areas; and advocates stewardship and conservation of natural resources.

DNS, headquartered at Ashland Nature Center, owns or manages more than 2000 acres of land for wildlife habitat and education preserves. Coverdale Farm is utilized for farm education and Community Supported Agriculture programs. Abbott's Mill's historic, water-powered gristmill is preserved and operational. The Society's Burrows Run and Flint Woods Preserves in New Castle County and Marvel Saltmarsh and Cedar Bog Preserves in Sussex County provide extensive field study opportunities. DNS operates the DuPont Environmental Education Center at the Russell W. Peterson Urban Wildlife Refuge on the Wilmington Riverfront and offers programming at Cooch-Dayett Mills in Newark and historic Buena Vista in New Castle. Unlike many nature centers with environmental education programs only, the Delaware Nature Society also has Natural Resource Conservation and Advocacy components

The Delaware Nature Society publishes books on natural history topics to increase interest in the region's wealth of plants, animals, and living communities and the need for their conservation. *Natural History of Delmarva Dragonflies and Damselflies* was undertaken to disseminate information about these fascinating insects and to encourage field exploration of the peninsula's fauna. Through this book the society seeks to heighten awareness of the value of all native species and the importance of protecting habitats and preserving natural areas.

The Delaware Nature Society annually offers programs on flora and fauna for children, adults, and families. For more information about programs, publications, and services, please contact the society by mail, telephone, or e-mail or visit the Web site:

> Delaware Nature Society
> P.O. Box 700
> Hockessin, DE 19707
>
> dnsinfo@delawarenaturesociety.org
> www.delawarenaturesociety.org
> 302.239.2334

About the author

Harold (Hal) B. White, III, is a Professor of Biochemistry in the Department of Chemistry and Biochemistry at the University of Delaware where he has worked since 1971 and has conducted research on enzymes, molecular evolution, and vitamin-binding proteins in eggs. He is nationally known as a biochemistry educator; however, his interest in insects in general and dragonflies in particular preceded his interest in biochemistry and education. He has been fascinated with dragonflies since he was a high school student in central Pennsylvania. This serious hobby has taken him from Canada to Mexico and across the United States. Hal has published articles on the dragonflies and damselflies of Delaware, Pennsylvania, New Hampshire, Massachusetts, and Acadia National Park in Maine; and he coauthored descriptions of the larvae of the Ringed Boghaunter and the Pygmy Snaketail, two rare dragonflies. From 1982 to 1990 he was Corresponding Secretary of the American Entomological Society, and from 1990-1994 he was Vice President. He currently serves as chairperson of the Society's education committee. Hal's website at the University of Delaware is: http://www.udel.edu/chem/white/.